CONTRIBUTIONS TO THE PSYCHOBIOLOGY OF AGING

CONTRIBUTIONS

AUTHORS: John E. Anderson, Jr., Ph.D.
Howard J. Curtis, Ph.D.
William E. Henry, Ph.D.
Lissy F. Jarvik, Ph.D., M.D.
Robert Kastenbaum, Ph.D.
Walter G. Klopfer, Ph.D.
George A. Sacher, Ph.D.
Harry Sobel, Ph.D.
Roy L. Walford, M.D.

TO THE PSYCHO-BIOLOGY OF AGING

Edited by
ROBERT KASTENBAUM, Ph.D.

Springer Science+Business Media, LLC

ISBN 978-3-662-38911-9   ISBN 978-3-662-39847-0 (eBook)
DOI 10.1007/978-3-662-39847-0

Copyright ©1965 Springer Science+Business Media New York
Originally published by Springer Publishing Company, Inc. in 1965
Softcover reprint of the hardcover 1st edition 1965

Library of Congress Catalog Card Number: 65-26950

Authors

John E. Anderson, Jr., Ph.D. Dean, Columbus College, Columbus, Georgia.

Howard J. Curtis, Ph.D. Chairman, Brookhaven National Laboratory, Upton, Long Island, New York.

William E. Henry, Ph.D. Professor of Psychology and Human Development, University of Chicago, Chicago, Illinois.

Lissy F. Jarvik, Ph.D., M.D. Associate Research Scientist, Department of Medical Genetics, N.Y. State Psychiatric Institute; Assistant Clinical Professor, Department of Psychiatry, Columbia University, New York, N.Y.

Robert Kastenbaum, Ph.D. Director of Psychological Research, Cushing Hospital, Framingham, Massachusetts; Lecturer in Psychology, Clark University, Worcester, Massachusetts.

Walter G. Klopfer, Ph.D. Professor of Psychology; Portland State College, Portland, Oregon.

George A. Sacher, Ph.D. Division of Biological and Medical Research, Argonne National Laboratory, Argonne, Illinois.

Harry Sobel, Ph.D. Chief, Section on Aging Research, Veterans Administration Hospital, Sepulveda, California.

Roy L. Walford, M.D. Associate Professor, Department of Pathology, University of California, Los Angeles.

PREFACE

This small volume offers current thinking and research that, hopefully, might contribute to the development of an adequate psychobiology of aging. It is probable that most gerontologists share the viewpoint expressed by Birren which considers the individual as ". . . a biological, psychological, and social constellation moving forward in time."* In practice, however, it is difficult to integrate the biological, psychological, and social aspects of the individual into a single conceptual framework. Investigators encounter formidable challenges when they remain on a single level of phenomena, *e.g.*, changes in cell populations, or changing patterns of interaction between the aging person and others in his social environment. How much more difficult the task becomes should one attempt to relate changes which occur on the cellular level to those which occur on the social level! Furthermore, close attention dissolves both "the cellular level" and "the social level" into a variety of sublevels that are easier to take apart empirically than to put back together conceptually. And we have not even mentioned

* James E. Birren. *The psychology of aging.* Englewood Cliffs, New Jersey: Prentice-Hall, 1964. P. 1.

the variety of other focal points implied in a psychobiological approach.

Progress in two directions is required: progress in scientific knowledge of each level of phenomena on its own terms (*e.g.*, what happens to the lymphocyte as its owner "grows old"?); and progress in our ability to conceptualize with the peculiar combination of freedom and rigor that is necessary to link disparate phenomena (*e.g.*, changes in lymphocyte populations with the individual's mental functioning and, possibly, with his longevity).

Each chapter in this volume attempts to contribute to one or both directions of progress. Some chapters maintain a steady focus upon a single realm of phenomena, while others move away from their "home base" to explore possible interrelations at other levels. There is one important similarity: each report has emerged from a systematic approach which incorporates a conceptual framework and empirical methodology. Here is not a sampling of "one-shot" studies, but a view of programmatic inquiries at various stages in their development.

Part I offers four psychological or psychosocial perspectives on aging. The relatively new developmental-field approach is set alongside a further contribution to disengagement theory (an "old-timer" in social gerontology, as it was first published in 1960). While developmental-field concepts have been involved in several lines of previous investigation, the present chapter is the first attempt at an essentially theoretical exposition. The concepts of *engrossment* and *perspective* are introduced here, and called upon in the interpretation of past research and the formulation of new hypotheses. Henry's chapter examines the original *engagement* and *disengagement* conceptualizations in the light of subsequent research. This discussion enriches the earlier formulations and further emphasizes the developmental orientation of disengagement theory.

Interpersonal theories of human behavior have become of major importance in the social sciences. Klopfer attempts to point the way toward an interpersonal theory of aging. Threats to self-esteem in later life are considered as well as those adaptive techniques most frequently used by the aged. Learning theory, that backbone of experimental psychology, has not been as conspicuous a contributor to theoretical gerontology as one might have expected. Anderson explores this situation in a spirit that may not endear him to traditionalists, but is likely to invite fresh attention to the import of learning theory for gerontology (and vice versa).

Part II offers five chapters from the biological perspective. Sobel addresses himself to one of the truly fundamental questions: "... whether non-reproducing cells have attributes of potential immortality or whether time alone imposes self-limitations ... do the causes of death lie within or outside the cells?" Implications of interest to social scientists are discussed.

Somatic mutation and immunological theories of aging are discussed by Curtis and Walford. Curtis advances the somatic mutation theory with results from a new experimental approach and consideration of the most recent findings reported by other investigators. Walford presents the case for an immunological theory of aging which is susceptible to direct experimentation, as illustrated by his own recent studies. Jarvik's inquiry into chromosomal changes with aging provides material relevant to the chapters noted above, and introduces other significant considerations, including relationships between heredity, intellectual functioning, and survival.

Sacher's analysis of brain structure and longevity is perhaps the most fully realized psychobiological approach of those presented here. His "behavior-oriented approach to longevity and aging" provides an instructive example of how one can work productively with both concrete phenomena and abstract conceptualizations simultaneously.

The present book might have been titled *The Biopsychology of Aging* in fairness to the biological contributors who have, in the present writings, reached further into the psyche than their psychosocial colleagues (this writer included) have gestured toward the physical workings of the human organism. All the approaches in Part I have their implications for the biological realm. However, the authors were confronted with the more immediate task of clarifying and advancing viewpoints that have either been newly created or newly applied to gerontologic problems.

The juxtaposition of psychological and biological presentations does not automatically create a psychobiology of aging. But a sense of "perhaps-we're-getting-someplace" was evident when two parallel symposia, psychological and biological, were presented at the 1964 annual meetings of the American Psychological Association. This was the first occasion known to the editor in which a representative set of psychological and biological theories of aging were presented as a unit; in a way it constituted the first "official" recognition of the significance of theoretical activity in gerontology. These papers were revised by their authors and comprise the contents of this book which, we hope, will encourage the psychosociologists and the biologists to venture into each other's "turf" as well as examine recent developments in their own spheres. Perhaps in the reader's mental laboratory a psychobiology of aging will take shape, assisted to some extent by the materials offered here.

The Division of Maturity and Old Age of the American Psychological Association sponsored the initial presentation of these papers; therefore, it seems appropriate not only to express verbal appreciation to the officers of this scientific organization, but also to contribute all royalties to the Division for the continued pursuit of its scholarly activities.

ROBERT KASTENBAUM

Framingham, Massachusetts
June 16, 1965

CONTENTS

PART I.
Psychological Perspectives

1. Engrossment and perspective in later life: A developmental-field approach. *Robert Kastenbaum*   3
2. Engagement and disengagement: Toward a theory of adult development. *William E. Henry*   19
3. The interpersonal theory of adjustment. *Walter G. Klopfer*   37
4. The import of learning theory for gerontology. *John E. Anderson, Jr.*   45

PART II.
Biological Perspectives

5. Aging theory: Cellular and extracellular modalities. *Harry Sobel*   57
6. The somatic mutation theory of aging. *Howard J. Curtis*   69
7. Immunology and aging. *Roy L. Walford*   81
8. Chromosomal changes and aging. *Lissy F. Jarvik*   87
9. On longevity regarded as an organized behavior: The role of brain structure. *George A. Sacher*   99

PART I
Psychological Perspectives

ENGROSSMENT AND PERSPECTIVE IN LATER LIFE:
A Developmental-Field Approach

1

Robert Kastenbaum

Aging is a process that progressively debilitates the human organism.

Aging is a process that adds valuable new dimensions to human life.

Aging is a process that alters some aspects of the individual's total functioning and leaves other aspects relatively unchanged, with the particular kind of alteration varying from one person to another.

Which of these definitions is correct? Perhaps it depends upon what particular elderly person we happen to be observing at the moment. There is a tremendous variety of behavior and experience in later life, as is the case at other age levels. Yet when it comes to formulating theoretical constructions, often we are tempted to regard aging as fundamentally a single-minded process in the psychosocial realm as well as the biological. Individual differences may be acknowledged, but attention is riveted to the assumed general process of aging.

The material presented here is based upon work supported in part by a NIMH postdoctoral research fellowship, USPHS grant MY-4818, and, currently, by USPHS grant MHO-1520.

A balanced view of aging would take into account both general and individual patterns. In fact, it is difficult to see how either type of pattern could be defined and understood without serious attention to both. This chapter will explore some of the possible interrelations between general and individual patterns of aging within the framework of developmental-field theory. Developmental psychology has concerned itself largely with *general* patterns of development in the *early* years of life. Here we will explore its potential for contributing to our understanding of *individual differences* in the *later* years of life.

Developmental-field theory

The developmental approach cannot be contained within any one particular theory or any one particular discipline. Primarily, it is a way of thinking about phenomena, whether the phenomena of music, mathematics, interpersonal relations, or the life of a single individual. Perhaps we are thinking developmentally whenever we are concerned with structure and function as temporal events.

Concepts of development are incorporated into many approaches in psychology and related fields, including psychoanalytic and disengagement theories. In addition there are what might be described as the "pure" developmental theories such as those of Piaget (1960) and Werner (1957). To have before us a general statement of how a developmental theorist regards human behavior, it may be useful to consider this summary passage by Anderson:

The individual is regarded as ". . . an open system with a very high rate of interchange within the system and with the environment in which it exists. The processes which evolve out of this interchange are irreversible. When we first study it the system is already a going and active concern upon which events and stimulation are superimposed. This implies that the system is itself in some degree a determiner of its experience. With age this system grows in size and differentiates . . . the entire process of development is inherently selective, and movement is in general from versatility to efficiency. In succession choice points are encountered at which the decisions are made which determine the ensuing developmental process.

"Through the basic mechanism of learning . . . the system acquires patterns of skill and knowledge. In the early period with its curiosity and manipulation, tens of thousands of specific reactions to stimuli are established; later, and to some degree simultaneously, these are integrated and organized into larger units . . . As the system moves along it passes through successive stages out of which come new properties and accomplishments to which subsequent periods of orientation are devoted.

"Pervading the whole and facilitating all the factors involved in development is the symbolic process by means of which the range of communication, control, and direction, as well as the emergence of new properties, is enormously increased" (Anderson, 1957, p. 44).

This characterization of developmental theory also implies a field approach. In fact it might be suggested that developmental and field approaches always mutually imply and require each other. It is not relevant to pursue this topic right now, but there is one consideration that will advance our main discussion.

Development is often regarded as a set of processes that occurs over a protracted period of time. The temporal unit may be in terms of months, years or decades. Field dynamics is often regarded as a set of interactions occurring simultaneously or within a relatively limited period of time in which the temporal unit may be on the order of seconds, minutes, or hours. Yet both developmental and field processes can be regarded as occurring in *either* a limited or an extensive temporal framework. Werner (1957) and Flavell and Draguns (1957) have described developmental processes that occur so rapidly that the situational rather than the longitudinal framework appears appropriate. The term *microgenesis* has been applied to these developmental phenomena.

If developmental processes can be observed with stop-watch in hand, then it also seems to be the case that field dynamics can be studied with reference to a calendar. The research area loosely designated as "time perspective" has been much influenced by Lewin's efforts to add long-range temporality to conceptualization of the life-space. In recent years such concepts as "subjective time-fields" have been proposed to encourage consideration of the ways in which the individual's own system for anticipating the future and interpreting the past operates as a dynamic field with long-range consequences (Kastenbaum, 1959).

What follows then is a developmental-field approach in which both longitudinal and situational frameworks are explored, with particular attention right now to patterns of behavior and experience in later life. It is hoped that this approach can contribute to our understanding of the fundamental similarities of people who seem to grow old in different ways, without the necessity to regard the individual differences as somehow unreal or unimportant.

Engrossment and perspective

Our first challenge is to find a theoretical focus that will draw upon many of the resources in the developmental-field approach and yet not burden us with the entire system of hypotheses and findings while we work within the limited scope of the present chapter. At this point we would like to introduce two concepts that might serve this purpose, the concepts of engrossment and perspective.

Consider a really satisfying kiss. When one is an active participant in a "good kiss," he may be said to be *engrossed*. While the lover is a-loving he is completely involved in a here-and-now transaction. At the instant of mutual engrossment it is as though both partners existed not as separate self-systems but as a single point of interaction. But the psychological situation is quite different when we *consider* the kiss. In attempting to relate the kiss to something else—to other kisses perhaps, or to the possible future consequences of the embrace —we are developing a perspective. If engrossment may be regarded as complete psychological involvement in one unitary situation, then perspective may be regarded as simultaneous involvement at two or more points in the life-space. The quality of engrossment vanishes when we compare, judge, plan, seek to explain.

Let me share with you now some of my basic assumptions concerning engrossment and perspective. These statements and some others which follow will not be adorned with the usual scholarly vocabulary of reservations, qualifications, and half-apologies, because space does not permit such a luxury.

First, engrossment and perspective refer to the operation of two basic psychological processes both of which are required for mature human functioning.

Second, these processes develop and function both genetically and microgenetically.

Third, many of the phenomena associated with aging can be understood as individual solutions in terms of engrossment and perspective to the onset of crises and depletions.

One way to become better acquainted with these concepts is through a brief survey of the development of time perspective from infancy through the adult years. This survey is itself a theoretical construction based upon a variety of observations by a number of investigators;[1] here we will be concerned only with those developmental aspects of time perspective that have the most direct bearing on the concepts of engrossment and perspective.

The infant is thoroughly engrossed in his immediate situation. Yet he seems to appreciate the difference between "presence" and "absence." This awareness, however, does not distinguish between temporal and spatial dimensions. Either something is both here-*and*-now, or it is absent, and "absence" perhaps is equivalent to non-existence. The adult, by contrast, differentiates "absence" into several temporal and spatial alternatives: something exists *now*, but not *here*, in this space; or, something will be or has been in this space, but is not at this time; or, again, something is neither here nor there, now or ever.

As development proceeds, the individual increasingly differentiates between temporal and spatial schemata, and makes further elaborations within each system. Appreciation of what is *not* here-and-now provides an alternative to being thoroughly engrossed in the immediate situation. Thus, the discovery that there really is a "future time" permits the child

---

[1] Recent reviews of the time perspective literature have been made by Fraisse (1963), Wallace and Rabin (1960), and Kastenbaum (1964c).

to escape partially from the domination of his immediate situation. Erikson's concept of "basic trust" (Erikson, 1959), and Singer's refinement of the concept of "capacity to delay gratification" (Singer, 1955) suggest some of the implications of this expansion beyond the momentary psychological field.

Expansion of the temporal range seems to increase through childhood and adolescence, although this is by no means the only development that occurs in the realm of psychological time. However, extension of thought into the future and/or the past is not identical with the construction of a time perspective. A person who is "wrapped up" in the future or is "living in the past" is still a person who is engrossed, even though this engrossment is not in the consensual here-and-now situation.

As indicated earlier, the simplest form of perspective requires that one point of attention be seen in relationship to another point. There is the further implication that the individual would be locating "himself" at one of these points and utilizing the other reference point to achieve the effect of perspective. Since it is temporal perspective that we are using for an illustration, one simple form would consist of evaluating the immediate situation in terms of its possible future consequences. A more elaborate perspective might consist of evaluating the immediate situation in terms both of past and future circumstances. More elaborate still would be a perspective in which there is a flexible shifting of emphasis among past, present, and future standpoints, with all three orientations always involved but varying in relationship to each other. Thus, for example, the individual may momentarily be locating himself in the past and thus call upon his immediate observations and scanning of the future to provide a context of meaning around a past event. At another time he may locate himself in the future and be concerned with planning for this future self by enlisting the other reference points for this purpose. Analysis of perspective thus requires a determination of where the individual locates himself in his own "temporal life-space" or "life-Gestalt," and how he is using other points of reference to organize a perspective.[2]

Before adolescence is completed, at least in our society, most individuals seem to have at their disposal all the necessary formal components for a genuine time perspective. The *integration* of these components into a meaningful personal perspective is one of the major developmental tasks not only of adolescence but of the succeeding years as well. I have suggested elsewhere (Kastenbaum, 1964c) that "Time perspective appears to serve two major functions: (1) liberates the individual from dominance by his immediate concrete situation, and as a corollary, provides an alternative to impulsive action; (2) provides a framework within which self-identity develops, maintains, and transforms itself." Furthermore (Kastenbaum, 1964b), that ". . . the extent and type of subsequent psychological development is highly structured by the kind of integration that is created fairly early in life, probably around late adolescence and the next few years. This integration will weigh heavily in determining whether or not there is to *be* further development, moreover, . . . in what particular pathway (subsequent development is likely to proceed)."

There is reason to believe that somewhere around the fourth decade of life many individuals enter a new period of psychological transition. Here there is a conflict between

---

[2] This discussion omits consideration of time perspective within an impersonal context, which is touched upon elsewhere (Kastenbaum, 1963; Kastenbaum, in press).

maintaining the previously established perspective and attempting to develop a new perspective that might accord more satisfactorily with the individual's present and future situation. This period is often characterized by explorations and vacillations, and might be described as a "flip-flop zone" within the total life-Gestalt. Often some degree of stability is then achieved and maintained for the next two decades. Another period of identity crisis is then a distinct possibility, especially if the earlier solutions at adolescence and mid-life have not been adequate. We will return to later life after considering a few points that should be made more explicit.

First, it should be noted that the concept of perspective is being used in two different ways: as an active process that develops and functions within the situational context, and as a more or less stable psychological structure that develops over a period of years. All the interrelationships between these two formulations cannot be explored here, but some considerations are quite relevant to the present topic.

The process of putting an event into perspective can be regarded as part of the "microgenesis of daily life"—we put our experiences into perspective many times in the course of a day, even within an hour. We also experience many engrossments in the course of daily life. If we could carry an impartial observer inside our head over the period of a few hours he would, under ordinary life circumstances, record numerous shiftings from engrossment to perspective and vice versa. Hypothetically, at least, the patterning of engrossments and perspectives could be charted as thoroughly as Roger Barker and his colleagues have charted the individual's external behaviors within his external behavior-settings (*e.g.*, Barker & Wright, 1954). On this view then, our daily life is made up of a more or less rapid succession of engrossments and perspectives. At times, however, we linger in a given perspective or engrossment. Some of us have individual dispositions toward lingering in an engrossed footing; others tend to linger in a perspective footing.

As one general hypothesis, I suggest that as a person develops he becomes increasingly adept in *shifting* between engrossment and perspective. The general integration of his personality is sufficiently secure to permit him to move from one footing to the other and still remain essentially "the same person."

Perspective may also be regarded as a more or less stable psychological structure, a pattern for the organization of experience that has been developed over a period of time and can be "tried on for size" when we are attempting to make sense out of a particular situation. Each individual has a number of such perspectives—and I suggest here a second general hypothesis, namely, that the number, variety, and scope of time perspectives increases with the advancement of the over-all developmental process. The mature individual, for example, can employ a temporal framework that includes his entire life-span, and also a variety of more limited frameworks, *e.g.*, from the last election to the next election, or from yesterday to tomorrow. As a corollary, there is an increased ability to shift from one perspective to another, as well as from perspective to engrossment.

So far it has been suggested that perspective functions both as a microgenetic process and a relatively stable psychological framework for the organization of experience. However, engrossment has been characterized only as a microgenetic or situational process. It is more difficult to do justice to engrossment both as a momentary phenomenon and as an enduring aspect of the total personality. In this chapter, the concept of engrossment will be limited to an immediate state of thorough involvement. The relatively stable structures that support moments of engrossment might be characterized very imprecisely as predispositions toward action. Thus, a given individual may have a strong predisposition toward loving, but will be engrossed in loving only at intervals.

One additional point should be introduced before we return specifically to later life. Engrossment and perspective can be regarded as concepts that are intermediary between the level of clear-cut empirical observation and the level of abstract theoretical construction. Two of the most abstract concepts in developmental-field theory are those of *differentiation* and *hierarchic integration*. Thus, for example, Werner (1957) proposed an "orthogenetic principle" which states ". . . that wherever development occurs it proceeds from a state of relative globality and lack of differentiation to a state of increasing differentiation, articulation, and hierarchic integration" (p. 126).

In general, what I have been suggesting about engrossment and tendencies toward engrossment may be regarded as a somewhat more specific form of differentiation, and what I have been suggesting about perspective may be regarded as a somewhat more specific form of hierarchic integration. Engrossment and perspective are rather general theoretical constructions, to be sure, but provide a conceptual bridge to specific phenomena, particularly in the realm of psychological time.

Pathways of individual development in later life

Several broad hypotheses concerning pathways of individual development in later life will be sketched very briefly. These notions have emerged from research and clinical work conducted by my colleagues and myself, and, hopefully, will be tested and refined by work now in progress in the areas of time perspective, death, response to psychotropic drugs, and individual and milieu therapy. A number of the specific observations have already been reported elsewhere;[3] no attempt will be made to summarize this work here. This brief exposition must also neglect many theoretical considerations.

Let us first conjure up an image of the mature individual. In terms of what has already been suggested, he has developed many differentiated systems of interest and competence and at times will be engrossed in a number of different situations. He has also developed a number of perspectives which permit him to organize and reorganize his experiences. Furthermore, our mature person is able to shift appropriately between engrossments and perspectives. What happens in the later years of life?

I suggest that many individuals begin to lose a psychobiological "something" that for the moment we will simply designate as "X," for it is the *effect* of this hypothetical "X" that is of most direct relevance. (If we consider this "X" to be a psychobiological energy, then we will be close enough for the present purposes). The diminishment of "X" poses a challenge to the individual's total identity and pattern of functioning. He must maintain himself on a "limited budget," so to speak, and this often will require a psychological reorganization.

Several alternative solutions are possible when this general

---

[3] Research by the present writer and a number of his colleagues is included in various chapters of *New thoughts on old age*, R. Kastenbaum (Ed.), New York: Springer, 1964. See also Kastenbaum, Slater, & Aisenberg, 1964, and Kastenbaum 1964 (a), 1964 (d), and 1965.

problem is encountered. The genuine possibility of more than one kind of solution should be emphasized, because both traditional developmental theory with its concept of "regression," and role theory with its concept of "disengagement" tend to dwell upon a single, general pathway of change.

One solution is to maintain at all costs the most general perspective or integration that the individual has previously achieved during his long process of development. As a corollary, we might expect this kind of solution to be accompanied by a microgenetic emphasis upon perspective in contrast to engrossment. This solution tends to preserve the general framework of total identity, but at the cost of sacrificing specific engrossments and differentiations. The framework remains; the content is blurred and perhaps fused. In an extreme case this type of individual might be characterized as the "Grand Old Shell." He still looks like the man who won the Pulitzer Prize and the Boston Marathon. Yet the appearance is virtually all that remains, and cracks in the shell disclose glimpses of confusion and decay.

Another solution is to de-differentiate and reorganize one's life on a simpler level. Here one sacrifices the framework of total identity for relatively competent functioning in one or two provinces of what formerly constituted the complete self. This solution for some individuals will involve a concentration upon the physical self; for others it will involve a concentration within the realm of symbolism and abstractions; for still others it will involve concentration upon social or interpersonal transactions.

There is still another general kind of solution that is attempted by some aging individuals. I refer to the alternation and vacillation between higher and lower central organizations, or between engrossment in one or another of the differentiated systems. The identity crisis is closer to the surface in many such cases, and in extreme instances I have found it appropriate to speak of multiple personality as a rather desperate solution for some aged people who are beset by overwhelming problems both from within and without (Kastenbaum, 1964a).

Very briefly, I would like to touch upon the relationship between the present formulation and disengagement theory. Cumming has stated that "Disengagement probably begins sometime during middle life when certain changes of perception occur, of which the most important is probably an urgent new perception of the inevitability of death." This changed perception, concerned largely with time and death, is said to be followed by a process of *mutual* withdrawal between the aging individual and his society. This disengagement process is considered to be a significant aspect of *normal* aging.

By contrast, I regard this process as just one possible pathway within one general solution to the challenge of aging. The disengagement solution requires a changed perception of time and death around mid-life. But some people have a pattern of perspective and engrossment within which this crisis does not develop. Some people, for example, remain thoroughly engrossed in day-by-day living—they never develop a significant perspective; time and death have little emotional reality, and this orientation is not necessarily changed by increasing chronological age. There are other persons who have such a highly developed, integrated system of perspectives and engrossments that they have worked out their basic relationship to time and death long before mid-life.

Furthermore, of those individuals who do experience a changed outlook at mid-life, some follow other pathways than disengagement. As already indicated, for example, some people concentrate their remaining energy within the social sphere, a solution that sacrifices introspection and the maintenance of total identity.

I am suggesting that the basic problem for the aging individual is how to find meaning and satisfaction in life as he becomes increasingly less able to keep everything going at once. He can select one or several "parts" for sustained involvement; he can try to preserve the "whole" at the expense of the "parts," or he can vacillate between these solutions. "Disengagement" is one rather popular solution to this underlying problem.

I recognize that by attempting to include so many differentiated elements within this chapter I may have failed to provide an adequate integration, and I cannot yet invoke aging as my excuse. But I do hope that I have been able to share some notions that might eventually contribute to our understanding of human development through the life-span.

References

Anderson, J. E., Dynamics of development: Systems in process. In D. B. Harris (Ed.), *The concept of development*. Minneapolis: University of Minnesota Press, 1957, 25-48.

Barker, R. G., & Wright, H. F. *Midwest and its children*. Evanston, Ill.: Row, Peterson, 1954.

Erikson, E. H. *Identity and the life cycle*. New York: International University Press, 1959.

Flavell, J. H., & Draguns, J. A microgenetic approach to perception and thought. *Psychol. Bull.*, 1957, *54*, 197-217.

Fraisse, P. *The psychology of time*. New York: Harper & Row, 1963.

Kastenbaum, R. Time and death in adolescence. In H. Feifel (Ed.), *The meaning of death*. New York: McGraw-Hill, 1959, 99-113.

Kastenbaum, R. Cognitive and personal futurity in later life. *J. Indiv. Psychol.*, 1963, *19*, 216-222.

Kastenbaum, R. Multiple personality in later life—A developmental interpretation. *Geront.*, 1964 (a), *4*, 16-19.

Kastenbaum, R. Is old age the end of development? In R. Kastenbaum (Ed.), *New thoughts on old age*. New York: Springer, 1964 (b).

Kastenbaum, R. The structure and function of time perspective. *J. Psychol. Res.* (India), 1964 (c), *8*, 1-11.

Kastenbaum, R. The realm of death: An emerging area in psychology. Paper read at 1964 (d) American Psychological Association meetings, Los Angeles, California. Revised version, *J. Human Relations*, in press.

Kastenbaum, R. The direction of time. I: The influence of affective set, *J. gen. Psychol.*, in press.

Kastenbaum, R., Slater, P. E., & Aisenberg, R. Toward a conceptual model of geriatric psychopharmacology: An experiment with thioridazine and dextro-amphetamine. *Geront.*, 1964, *4*, 68-71.

Kastenbaum, R. Wine, pleasure, and aging: An exploratory action program. *J. Human Relations*, in press, 1965.

Piaget, J. *The psychology of intelligence*. New York: Littlefield, 1960.

Singer, J. L. Delayed gratification and ego development: Implications for clinical and experimental research. *J. consult. Psychol.*, 1955, *19*, 259-266.

Wallace, M., & Rabin, A. I. Temporal experience. *Psychol. Bull.*, 1960, *57*, 213-236.

Werner, H. The concept of development from a comparative and organismic point of view. In D. B. Harris (Ed.), *The concept of development*. Minneapolis: University of Minnesota Press, 1957, 125-148.

Werner, H. *Comparative psychology of mental development* (Rev. Ed.), New York: International University Press, 1957.

ENGAGEMENT AND DISENGAGEMENT: Toward a theory of adult development

2

William E. Henry

Disengagement: concepts and data

In an age of anxiety, detachment seems an inappropriate art, and activity a solution. But wisdom and perspective—so commonly, though conceivably erroneously, attributed to the elders—require a base in leisure. By the term *leisure* I do not mean the common American association to *recreation*, that is, the *utilization* of leisure in particular activities. I refer, rather, to the root definitions of leisure as given in *Webster's New International Dictionary,* definition 2, "freedom or opportunity afforded by exemption from occupation or business" and, particularly, to definition 3, "time at one's command, free from engagement." The crucial ideas here are contained in definition 3, and they suggest two important concepts. The first, "time at one's command," I take to mean that the individual is able to structure time *as he chooses,* or can *ignore time altogether,* since it is he who commands it, rather than time that commands him. The second idea, "free from engagement," implies a possible logic for that kind of choice,

---

Paper read at the American Psychological Association, Los Angeles, 1964, adapted from a paper read at the International Gerontological Research Seminar, Markaryd, Sweden, 1963.

that is, that the individual is *free from the sense of being engaged* by various outer events or objects. Thus when the individual is not bound, i.e., not engaged with events and objects, time alters its relationship to the individual, and the individual alters his view of the importance of being engaged. Objects become less vital and time loses its ability to command.

A conviction that time is important and engagement with persons and objects necessary constitutes a firm basis for the maximal involvement with outer world events characteristic of middle-age (and possibly middle-class) life. The raising and training of families, the accumulation of personal property, the commitment to social and occupational mobility are in no fundamental personal sense inevitable. They may be *necessary* for the society, as mechanisms of inducting the young and as means of getting the world's work done. And they are perfectly *natural*, in the sense that they are, in most societies, common or modal events. But to become functional for the individual, they require the commitment of engagement. Once the individual to some degree *commits* himself to such values, the activities and objects related to their attainment constitute the interactive scene of early and middle adult life.

But for some adults such interactive values and activities become impossible, and for some possible but meaningless, except insofar as these persons come to invest energy and belief in the outer events and objects characteristic of the adult life scene in their society. This investment is the process intervening between individuals and social actions. It includes intertwining and probably inseparable elements of *commitment to time, and cathexes to objects, persons, and events.* This process, including the commitment of energy and the assumption that time can command, constitutes the necessary basic condition for the engagement in social activities at certain life periods. It is the dissolution, or perhaps only the reassignment, of the energies involved in this process that constitutes the basic condition for the disengagement of later life.

In the report of Elaine Cumming and myself in the book *Growing Old* (1961), based upon data from the Kansas City Studies, we emphasized the concept of *disengagement* as useful in accounting for the reduction in life activities and in ego energy found in most, but not all, older persons. It was our proposal that this was a process both social and psychological and, as such, to be explained as much by psychological events indigenous to the individual as by societal reactions relating to the older persons' inclusion in, or exclusion from, the work and social life of the society.

In the report of Havighurst and his colleagues, given at the Sweden Conference and based upon additional material but from a somewhat depleted sample of these cases, the disengagement process is further examined and the relations of disengagement, activity-maintenance and life satisfaction are more intimately studied (Havighurst, Neugarten & Tobin, 1963). A crucial part of these more detailed findings deals with the fact that the fate, in old age, of the cathexes of an earlier age, appears to depend upon the character of the personality coping mechanisms. While it is clear that varying life circumstances, social and biological, influence the psychic choices made, the manner of resolution and the techniques of coping begin to emerge as crucial differentiating factors. It is in this sense that *styles* of aging are composed of differing patterns of role activity and by differing patterns of relation to self. When Havighurst, Neugarten, and Tobin comment that the individual, regardless of the nature of his social life, does not *disengage from self,* they have made a most important statement. The nature of that self then becomes the crucial variable. A similar conclusion would appear to be implied by Williams and Wirths (1963) when they make productive differentiations in *lifestyle* by utilizing the essentially *self*-relevant categorizing variables of autonomous-dependent and persistent-precarious.

Disengagement theory, as initially phrased, proposes in rough outline a severing of ties between a person and others in his society, a reduction in available ego energy, and a change in the quality of those ties remaining. It proposes that the changing quality of remaining ties may stem from "an altered basis in the person for the reception and initiation of social events" (Cumming & Henry, op. cit., p. 107). We have suggested that this changed basis in the person resides essentially in a realignment of the relation of inner events to outer events in such a manner that the former take on an increasing centrality, that interiority becomes increasingly important. It is consonant with this that normatively governed considerations and hierarchally structured social relations should be less prominent. The proposal of the increased role of inner events in the reception and initiation of social events also permits the observation, variously made in other contexts, that the older person becomes increasingly "like himself," that, as Havighurst, Neugarten, and Tobin comment, he remains engaged with self. If these are reasonable observations, then they are so to the extent that new adjustments to changing norms become less common. The individual thus does not become more like the social scene, changed or unaltered as it may be. He becomes more like his own inner events, the sameness and continuity of which now have every opportunity to be maximized. This would presumably still be true even if, as appears commonly to be the case, general available ego energy declines. The issue here is on what basis personal or social events are judged, what social activities mean for the individual, regardless of whether these events and activities are many or few.

If overt work activity, or any social engagement, has meaning for the individual, that meaning probably is best described through relation to personality concepts, to social interaction values, or both. The personality coping mechanisms seen by Havighurst, Neugarten, and Tobin as crucial in accounting for many of their findings, would seem to me to constitute evidence for an "altered basis in the person for the reception and initiation of social events," of which I have spoken. In other words, these findings considerably advance and modify this aspect of disengagement theory, by giving us a more specific image of the nature of the inner events characteristic of the elderly. That these various personality coping mechanisms are accompanied by differing life patterns and by a differing sense of life satisfaction would seem to suggest that they do indeed serve as a screen through which outer events are seen and judged.

In a broad sense, a work task for a ten year old may take its essential meaning from the affect relationship to the father it may embody. A work task for the 30 year old male may serve to prove his instrumentality and his independence of that same father. For the 45 year old it may serve also to demonstrate his social stability, his responsibility, and his care for his family. And for a 64 year old (assuming retirement at 65) work may be related to efforts to prove to himself he is worthy of continuing work, or become merely a marking of time—something he must do while he waits for retirement.

If the meaning of work does indeed alter with age and social circumstance in some such manner as this, then we would add, for the post-64 year old, a further stage. This stage would be one in which there occurs a change from the dominance of *instrumentality* to that of *socio-emotionality*. It is in this sense that the suggested increased centrality of self-oriented interiority comes into play and that rewards for activity become dominantly judged in terms of their gain for self, rather than for the gain they might engender for that individual in the work or social scene.

Persons of advanced age who maintain high activity, in contrast to the common pattern of reduction in activity and ego energy, are cases of special interest. Conceivably they are, and always have been, biological specimens of great vigor. Conceivably they are persons, both uncommon and special, whose personal and social circumstances have never presented them with the choice of disengagement and who find the continuance of high activity meaningful and satisfying. But this latter statement is more a re-statement of the fact that such persons exist than an explanation of them.

In our statement of disengagement theory, we said that the loss of certain central roles will result in a loss of morale *"unless different roles, appropriate to the disengaged state, are available."* (Cumming & Henry, op. cit., p. 215). On the face of it, these are persons who have found some alternative roles. They are persons for whom the term "re-engagement" or "re-organizers" of roles may indeed be fitting. The emphasis in these terms (re-engagement, re-organizer) upon a dis-engaged period followed by a re-engagement cannot be clarified from our present Kansas City analysis. Neither is it clear whether (rather than dis-engage and then re-establish contact) these people have merely re-grouped or re-organized their life activities into a new but still high activity pattern.

Certainly either or both would be possible. But regardless of these possibilities, another issue suggested by disengagement theory requires resolution. In our statement quoted above, it is noted that different roles, appropriate to the disengaged state, may be adopted. The question now becomes whether or not the activities of the older period for these high activity persons are "appropriate to the disengaged state," and, further, what the more precise meaning of this somewhat ambiguous concept is. First, let us agree that some persons may well maintain high activity, with good morale, where those activities are indistinguishable from the clearly instrumental roles of middle-aged life. There are those among the Kansas City group who fit that pattern. It may be that the Reichard-Livson-Peterson (1962) *Mature* men fit that pattern. However, while these authors appear to identify only their *Rocking Chair* type with the notion of disengagement, it should be noted that their reports of the *Mature* men include a great deal of socio-emotionality, sensuality, direct expression of enjoyment in low-activity—while clearly maintaining interest in some activity, hobbies, and other part-time substitutive instrumentality.

In fact, their quotation from one of their mature men, presented in evidence of his active interest as an "enthusiastic breeder of chinchillas" (p. 119), strongly suggests some of the properties of the disengaged, post-64 age state and a rather uncommonly sensual one at that. The full quotation given is:

> "I saw one of the most beautiful chinchilla herds: I didn't know they existed. This fellow had been in it eight years. And he has the most beautiful animals. I could just sit there and watch them all day. And I was glad I went down to see him. It gives me something to work for."

It is of course inappropriate to imply without fuller material, that this man is in any particular stage of disengagement, Reichard and her co-authors addressed themselves only peripherally to issues of disengagement, and there is little question that their *Mature* group is a meaningful one in its own terms.

The chinchilla breeder, however, in this quotation, strongly suggests that his ego energy has been displaced from active instrumentality and that the still active work involved in chinchilla raising serves him quite differently. Its meaning is not instrumentality in the usual sense, but rather an occasion for unassertive emotional experience and for sensuous, even voyeuristic, observation. Such an orientation, of course, in no way denies the possibility of active work with his chinchillas or even that he may not achieve some instrumental results—in the actual sale of the animals, for example. The basic issue is one of the meaning of the activity to the individual.

Cumming and Henry have also given illustrations in which overt impulse enjoyment seems to play a crucial role. It is indeed our impression that in some instances the release from normatively governed demands, the dropping of "socially expected instrumental tasks," as Williams and Wirths note, tends to be accompanied by a kind of sense of release in which such pleasurable sensuous elements are often prominent. This sense of release is perhaps behind many of the so-called eccentricities of the aged, in which, at least in an affluent society, both the aged and their younger contemporaries seem to take pleasure.

Focus upon interiority

But this form of expressiveness in disengagement may not be the most common, nor may it be even possible in some cultural contexts. *The basic issue is the resurgence of focus upon interiority and the release from the sense of import formerly attributed to outer world events.* It may well be that the factors determining the specific form of reactions at this point reside in the life style characteristic of the individual's life to that time, and in the characteristic coping mechanisms of the personality. For many Americans, leisure and freedom from occupation have commonly meant the opportunity to play, that is, to be active in non-work contexts, to "enjoy" oneself (on the assumption that one could not in work), to be amused (on the grounds that work was serious). It may be for this reason that "going fishing" is such a satisfactory symbolic, if not actual, activity as an expression of release from work. For many it is a classical form of enjoyment without occupation.

But for others the contemplation of philosophies or of nature may have constituted the definition of the higher, non-obligatory life, that life for which the individual previously was led to believe he had no time. These concerns should then become dominant as appropriate to the disengaged state.

Dr. Talmon-Garber, in personal conversation, has noted that many older persons in the Kibbutzim in Israel do not seek expressivity in the more playful manner just suggested but, rather, devote increased proportions of time to reading—largely of the Bible and other philosophic works. In a sense, this activity is a continuation, even an accentuation, of interest in certain values already a dominant part of their previous lives, values appropriate to their cultural setting. I should think that such socially sponsored and non-instrumental activities should also serve well the gains of interiority and self-contemplation. In a religion-oriented society the philosophic values documented in such reading presumably represent values against which the reader's own personal life had been judged in younger days; and a re-examination of them in later life would seem entirely fitting, both as a socially permitted activity and as a re-contemplation of internally-based, ego-relevant ideas. Certainly, also, the fact that neither active instrumentality nor overt social activity are required elements gives such reading a particular appropriateness.

For individuals whose lives have been highly involved with group and organizational participation, and for whom high value has been placed upon instrumentality in that context, one would suppose that their group and organizational scenes would remain prominent among later associations, even though there might be reduction in total amount of interaction. The crucial issue as to whether, despite level of activity, some disengagement had occurred, would reside in the *nature* of the activity at the later age. Disengagement theory would suggest that the activities maintained could be less instrumental, and that the instrumental ones would be the first dropped, with less official attendance perhaps remaining. For these people, the observation of Williams and Wirths that (for our Kansas City sample) television is "the great disengager," is amply relevant. American TV programming permits that the individual older person can, with complete physical and psychic immobility, contemplate and passively re-examine many values inherent in his earlier life and in the social system at large. The virtues, or at least the common attributes, of achievement motivation, of over-determined social engagement, of complex mutually-dependent relations, are all fully displayed. Their merits and their relation to self can easily be re-examined with no commitment whatsoever.

Our concern here is with the degree to which specific forms of mid-adult activities and values serve as the significant basis for judging the particular kind of activities crucial for any particular individual in old age. It would seem likely that the activities of mid-age are the key to the particular interactive scenes of old age. This position is consistent with the proposal that men whose mid-adult activities already stressed interiority and socio-emotionality (ministers, educators, for example) find the transition into lower instrumentality and higher socio-emotionality an easier one than men for whom the values and habits of interiority have always been minimal. In a broad sense, the values of the society in question—its definition of the good and the bad, of the permitted and non-permitted, of work and play—influence not only the timing of a possible disengagement, but the activities and interests which might serve functionally as appropriate to that period.

It would also seem fitting if specific forms of disengagement were effected by the degree of tenacity with which, during mid-adult life, the individual clung to his society's definition of the good and the proper and, in particular, the degree to which strong superego restraints influenced choices of action. In an important sense, strong superego restraints represent a flight from interiority, a strict guard against the consideration of actions directed by inner impulse and a preference for actions dictated by social convention, whether those of a contemporary scene or those of the remembered past. These persons should strongly resist the possible upsurge of interiority, while participating in societal activities to the degree that the conventions of their society demanded. For these persons, societal permission to disengage becomes crucial. They may find the adoption of an expressive and socio-emotional style difficult and threatening. These would not be the persons reported by Havighurst, Neugarten, and Tobin as high in both social activity and morale. They should rather be those who tended to remain engaged beyond a period of personal satisfaction in those activities, and to be low in morale—low because their fear of interiority permits them to envision no dignified alternative to the social modes of their mid-adult life style.

Engagement and involvement as developmental events

Earlier, in this chapter, I called attention to the possibility, and I believe it to be a fact, that *en*gagement and the derivation of a sense of commitment are in themselves developmental events. In at least some formulations the schizophrenic is the individual who, at an early age, and presumably through uniquely traumatic circumstances, has never developed the sense of involvement characteristic of most persons. The more common developmental pattern, of course, is the one in which psychic and material gains are the clear-cut rewards, and subsequently the motives, for a demonstrable degree of interest or investment in significant persons in the immediate environment. This process of developing reciprocal involvements is presumably facilitated by firm identifications with these figures and comes to diffuse to include actually and symbolically related persons and events. With increased ego differentiation and the development of autonomy in ego and in other spheres, there probably occurs a fairly stable level of involvement—one to some degree reflective of the past experiences of the individual. It would seem logical to assume that this level of involvement would become greater at some times than at others. For one thing, it perhaps increases at points of new interest—*e.g.*, at adolescence, and at periods of early development of work competence and of establishment of a family. These are points at which a fairly generalized sense of the import of all commitments becomes more acute. One could probably maintain that some increase in the level of involvement occurs at the time of each of what Havighurst has called the Developmental Tasks.

But in spite of the probability of alterations dictated by special circumstances—and certainly retreat from involvement is also possible—it seems worthwhile to posit a general personality characteristic residing in this level of involvement. The issue would be one of the psychic distance or closeness of the individual to the common objects and events of his environment. Closeness, in this sense, implies that the individual maintains a high awareness of the presence and the stimulation of others and that he himself experiences some sense of need to respond to that stimulation. Distance presumes a reduced awareness of such environmental events and a reduced felt need for response. Phrased somewhat differently, the presumption might be that individuals equilibrate themselves at some characteristic distance from others and, other things being equal, tend to maintain that distance. It does not follow from this notion of level of involvement that either closeness or distance carries with it any particular state of awareness of one's own inner states, though some relations of the two may be common for certain ages and/or for certain persons differing in other personal attributes. Thus, it would be consistent with this proposal that highly engaged persons in mid-adulthood would experience high involvement, with closeness to other persons and events, and that during this period formal attention to inner states declines. During classic periods of "negativism," that of the three year old American child, perhaps, and the "selfish" periods of adolescence, one might imagine a preoccupation with inner states to the exclusion of formal attention to involvement with environmental events.

The resulting distance from others could readily produce these accusations of "selfishness" and "negativism." It is, of course, also our suggestion that, as Cumming (1960) has said, "the older people are, the more likely they are to be equilibrated at a quite considerable distance from their fellows and the more preoccupied they are likely to be with their own inner states."

In this context, engagement and disengagement become a general form of personality dynamic, and the disengagement of the aged becomes a special case. Further study may suggest that some of the high or low engaged persons of older age are so in part because of previous developmental experiences indicative of particular levels of involvement. This would most likely be true of those well-known cases of extremely high engagement in old age whose earlier lives appear to have been similarly characterized. It seems intimated, also, by the remark of Williams and Wirths (1963b) that "preparation for disengagement" before 65 increases the likelihood of successful aging. While forms of "preparation" may well vary, one such prior event may be either an initially low involvement, or a particularly flexible sense of involvement, permitting ready adaptability to a retreating social scene.

Open questions

We said earlier that the disengagement of the aged is an intrinsic process and that it is inevitable. I would still be inclined to maintain that both those statements are true. However, its inevitability is by no means clear. There are Kansas City cases in whom signs of disengagement have not appeared at quite advanced ages, in at least one of age 86. There are certainly others in whom high activity is still present, but in which the question of altered personal meaning and the switch to increased dominance of socio-emotionality is not resolved. The question of inevitability, however, is less intriguing than the more refined questions variously phrased by the persons working on these data. The examinations reported by Havighurst, Neugarten, and Tobin revolve around the highly complex interactions of activity, ego energy, and the judged estimates of life satisfaction. From their work, it is apparent that several styles of aging are possible, and that the style itself is no criterion of felt satisfaction. Using a different concept of success and typologies based on a more socially interactive concept of life styles, Williams and Wirths derive similar conclusions, focussed somewhat more generally upon estimates of disengagement. From this point of view, disengagement occurs in all life styles, but is clearly subordinate, in its timing and nature, to some properties of those long-standing life styles.

The degree to which disengagement is an intrinsic process seems still an open question, and one related in part to how biologically-determined one assumes an intrinsic process to be, and where one places the causality in the various interrelations of social fact, personality, and self-estimate with which we have been dealing. I have earlier suggested that the process is intrinsic in the sense that it represents a personality process of level of involvement, and that that process is clearly a lifelong one. The emphasis which Havighurst-Neugarten-Tobin, and Reichard-Livson-Peterson place on personality as crucial determiners seems to support that notion—at least to the extent that the personality processes dealt with are presumably not specific to old age, but rather are based in earlier life experiences, and maintain considerable determining power in their interaction with the life events of old age. The possibility that evidences of ego-energy-decline precede evidences of reduction in social activity is similarly suggestive, though here problems of differential perception of actual changes confound the picture. To this point it would seem sound to me to suggest that processes of disengagement, in their complexity and interaction with various life events, are intrinsic in the sense only that social environmental events are not sufficient to predict them, and that they appear clearly related to various personality processes generally understood to be of long duration. They are thus ego processes, having a developmental history, a discoverable course of their own, and a positive power that can influence reaction to external events and choice of response to them.

But there is one further highly complex question which I think none of us working with these data have dealt with directly. That is the question of *intentional outside activation*, its utility in social planning, and its effect upon the aged. There are, of course, some implications for this in the present work. There is a presumption that low activity is good because we find examples of it in persons of good morale; there is a presumption that high activity is good when one finds examples of it in persons with high morale. In this context both the Activity Theory and the Disengagement Theory—acknowledgedly less distinct and separate than their names imply—carry the burden of their titles. But, as I think has been amply shown by the work reported, neither of them is sufficient to account for the facts we now know. And neither deals explicitly with the question of the effects upon the persons of specific plans for increasing or decreasing their activity, or for plans of providing one kind of activity as opposed to another. There are, of course, many personal reports of casual as well as intense professional work. As far as I am aware, these reports lend themselves to either or both general theories. And those that appear most substantially to support the implications of one or the other position have arisen from unsystematic circumstances without controls. I take this to be an open and challenging field of research, in social action and psychotherapeutic management. The present group of reports dealing with disengagement should provide some clues for a more refined experimentation and lead to further study of the interactions of personality and social events, and their meaning for life satisfaction.

References

Cumming, E. Social change and the dying process. Read at the 13th Annual Conference on Aging, University of Michigan, Ann Arbor, June, 1960.

Cumming, E., & Henry, W. *Growing old*. New York: Basic Books, 1961.

Havighurst, R. J., Neugarten, B. L., & Tobin, S. S. Disengagement and patterns of aging. Read at International Gerontological Research Seminar, Sweden, August, 1963.

Reichard, S., Livson, F., & Peterson, P. *Aging and Personality*. New York: John Wiley, 1962, pp. 119-120.

Williams, R., & Wirths, C. Styles of life and successful aging, II. Read at International Gerontological Research Seminar, Sweden, August, 1963 (a).

Williams, R., & Wirths, C. Some reflections on disengagement. Read at the International Congress of Gerontology, Copenhagen, August, 1963 (b).

Webster's New International Dictionary (2nd edition), 1959, p. 1414.

THE INTERPERSONAL THEORY OF ADJUSTMENT

3

Walter G. Klopfer

The originator of the group of theories usually labeled "interpersonal" is Harry Stack Sullivan who stated on many occasions that he considered personality in a vacuum as a meaningless concept and that it can only be studied operationally in an interpersonal context. Other theorists who followed him along this path were Erich Fromm, Erik H. Erikson, and, more recently, Timothy F. Leary (1957). The common element in all the concepts of adjustment by these theorists is an emphasis on the behavioral interaction between persons for two purposes. First, there is the alleviation of anxiety by improving interpersonal communication, and secondly, there is the mutual enhancement of sense of self-esteem by provoking positive feedback from others.

An appropriate sense of self-esteem is one characterized neither by excessive and unbridled grandiosity nor by inappropriately critical and self-derogating feelings. Anxiety is produced by any attack upon the reality of the sense of self-esteem by means of threats either from external frustration or unrecognized internal conflicts. In turn, anxiety is thought to require the use of defensive or adaptive techniques of various kinds in order to reduce discomfort and raise self-esteem back to the level of tolerance or endurance.

The present chapter will attempt to evolve an interpersonal theory of aging based upon the above-mentioned framework. First, we will review threats to the self-esteem that are indigenous to the aging period. Secondly, the adaptive techniques open to the aged individual to deal with these threats to the sense of self-esteem will be cited, and, finally, the worth of interpersonal theory for studying the aging process will be evaluated.

Threats to the self-esteem in the aged

1. The elephant herd in the jungle consists of a bull elephant, a number of cow elephants, and young ones of both sexes. Whenever one of the younger male elephants feels the urge, he lifts up his trunk and trumpets a challenge to the leader of the herd. The issue is that of combat to determine who shall subsequently be the leader. The two animals have very different feelings as they enter the fray. From the vantage point of the younger elephant, there is everything to gain and nothing to lose. If he should emerge the victor, the herd will be his and the former leader, dethroned and displaced, will have to stagger out into the bush to be seen no more. If the challenger should lose, he can merely wait until he grows somewhat stronger and try it again. The older elephant, on the other hand, has everything to lose and nothing to gain. Even if he should win, the victory will only be temporary, and he knows that the younger elephant will grow stronger as he grows weaker. If he should lose, it is the end.

This saga reflects a phenomenon true among humans as well as elephants. The phenomenon is that of the undeclared war which exists between one generation and the next, between youth and age. Young boys and girls envy the mature independence of their parents. When the children, in turn, become middle-aged, they assume the role of the dominant generation, their parents having in the meantime become old and dependent. The temptation to be dominant towards the elderly parents, and perhaps treat them with a lack of consideration provoked by hostility, is very great. Insofar as this situation is known and sensed by the aged, it represents a threat to their sense of self-esteem.

2. The society of the Western World tends to worship youth. Only young people are considered beautiful and desirable from every point of view. Women, in particular, make every effort to retain their youthful appearance by means of foundation garments, hair dye, pancake make-up, and many other means. Among men, too, it is considered socially desirable to appear young and vigorous long after this is really possible on the basis of reality. The attitude of society as such has tended to make all elderly people feel inferior, inadequate, ugly, and unattractive for no other reason than that they are old. This certainly constitutes an extremely important and debilitating narcissistic trauma to the aged. In no other society is it necessary to experience the feeling of guilt simply because of a condition over which no voluntary control is possible, namely, the aging process.

3. Older people, especially when they are forced to retire from their usual activities, experience decreasing sources of interpersonal satisfaction because of the dying off of old friends and lack of access to the peer group due to lack of employment. The man on the job has a built-in status. Even if he is on the lowest rung of the occupational ladder he still is the most senior member of the lowest rung. His seniority gives him some status as such. Even greater change is in store for the man who has achieved real status and importance in the world of affairs. Constant feeding of the sense of self-

esteem takes place while the individual is at work, and the sudden withdrawal of such satisfaction may cause severe threats to the sense of self-esteem.

4. Both physically and psychologically, older people lack resilience in recovering from stress or illness. When catastrophy strikes in the form of debilitating somatic symptoms, loss of property, or rejection by a significant other, the older person is traumatized much more than a younger. A man of 25 may lose his house, his job, or his wife and still have time to replace each one of them. The elderly individual would find any of these losses much more difficult to absorb, and is more likely to become despondent. Certainly, the catastrophic reaction to loss is another threat to the sense of self-esteem of the aged.

5. Finally, there are the obvious psychological sequelae of the various clinical conditions indigenous to the aged period. These include poor memory, loss of spatial orientation, loss of speed and agility, loss of sensory acuity, etc. All of these conditions represent handicaps. Insofar as they reduce competitive efficiency, and particularly insofar as they reduce the ability to communicate effectively, they will take their psychological toll. Sensory disability is likely to make the aged individual prone to suspicion and distrust of others since he has to use his imagination as to what is occurring in the world about him and is likely to suspect the worst.

Adaptive techniques employed by the aged

Disengagement. The theory of disengagement has been described by Henry (this volume). Certainly, disengagement is of value to some in dealing with their problems. In a study by Reichard, Livson, and Peterson (1962), alluded to in Henry's paper, one of the groups (the "rocking chair" group) was characterized as being able to use successfully the techniques of disengagement. In the opinion of the present writer, another group (the "self-haters") also can be considered as disengaging—but this time within the context of failure and maladjustment. The inference drawn here is that disengagement is an adaptive technique useful only to those who have the proclivity for it. There are individuals who have always been somewhat detached and whose engagement with others has been of necessity in order to meet the exigencies of the workaday world. Such individuals, once they have a social sanction for disengagement as part of their own decreasing involvement in the workaday world, can then happily proceed to disengage. However, those upon whom disengagement is foisted willy-nilly against their wishes and against their desires do not seem to react well to it.

Activity. In a study by the present writer conducted some years ago (Klopfer, 1958) it was discovered that active or productive people had generally more self-esteem, a more positive concept of the future, and a greater identification with children or child-surrogates. In a more recent study by Davis (1964), those aged subjects who were considered active or productive had generally more realistic aspirations and a better morale. Two of the groups in Reichard's study (1962) might be considered as activists. First, there were the members of the "armored group" who are the more successful activists; then there were the "angry men" who might be considered unsuccessful activists. Activity can take many forms. It might take the form of actual employment, religious activity, recreational activity, political activity, or club and volunteer activities. All of these may serve the same basic psychological function of enabling the person oriented towards productivity to continue producing in the only way that makes sense to him and thus retain his preferred self-concept.

Paranoid retreat. This is a frame of reference in which society is considered as the enemy, and life is one long battle of "one-upmanship." Feelings of frustration and hostility are projected onto the environment and attributed to others, or to society at large. Within the psyche of such aged individuals a process of selection may take place in which paranoid retreat is the only alternative to withdrawal and death. The question may revolve around either the turning of hostility inward, resulting in deep depression and possible suicidal tendencies, or the turning of hostility outward, resulting in a dysphoric and paranoid attitude toward society. Old people who are depressed see death as the only release from an intolerable situation. A paranoid stance may prove the only alternative in some instances, and, even though it may appear pathological, may still serve as a way of keeping self-esteem from becoming utterly destroyed.

Integration. This is implied by Reichard's "mature" group. Complete integration and adjustment in old age as a successful way of maintaining self-esteem and of relieving oneself of anxiety is, of course, very difficult to define. It probably includes some elements of all of the above. Thus there is some disengagement, withdrawal from competition enabling the aged individual to keep from suffering further narcissistic injuries. There may also be some continuous activity and construction occurring so as to enable the aged individual to receive some ego-supplies through this route. At the same time, some paranoid defenses may be used in order to ward off intolerable hurts, and depressive withdrawal may upon occasion be used to forestall further rejection. Such an individual would be able to age gracefully and accept aging instead of constantly trying to fight against it.

Conclusions

Interpersonal theorizing seems to be an important approach to the problem of aging since the major changes during the aged period, insofar as they constitute a subject matter of behavioral science, are in the interpersonal area. Many of the important practical questions in the area of aging are also in the interpersonal area. For example, should old people be integrated with younger people in institutional or other settings, or is segregation more desirable? Also, there is the question of whether old people should be placed in productive employment situations, or whether a recreational setting is more appropriate. Finally, there is the extremely important question of whether engagement or disengagement should be promoted by those influential in the field. The answers to these questions can be discovered eventually by studying them in an interpersonal context.

There seems no doubt on the basis of the evidence and the logic seems irrefutable that aging does constitute a narcissistic threat to almost everyone so "afflicted." Whether an aged person does or does not succeed in adjusting to this situation to the point where he can be reasonably content and satisfied would depend largely on whether or not he can master appropriate adaptive techniques to deal with the various threats to self-esteem cited above.

References

Davis, R. W. *Disengagement and the goal setting behavior of older people.* Unpublished doctoral dissertation, University of Portland, 1964.

Klopfer, W. G. Psychological stresses of old age. *Geriatrics,* 1958, *13,* 529-531.

Leary, T. F. *Interpersonal diagnosis of personality.* N.Y.: Ronald Press, 1957.

Reichard, Suzanne K., Livson, Florine, & Peterson, P. G. *Aging and personality.* New York: Wiley, 1962.

THE IMPORT OF LEARNING THEORY FOR GERONTOLOGY

4

John E. Anderson, Jr.

The knowledge-to-chaos ratio

The question of how learning occurs has been of interest to man from the earliest pages of recorded history down to the present time. From simplex explanations describing the process in terms of icons traversing the humoral paths in an organismic context of the unfolding entelechy, to complex ones describing the same process in terms of bits of information conveyed by sensory input mechanisms to hierarchical levels of central storage and processing assemblies in a black box context of a servo-mechanism, man has attempted to increase the knowledge-to-chaos ratio that inheres within the inquiry into nature and its processes. But early in this quest, man discovered that the *process* of inquiry was different, if not separate, from the nature of that into which he was inquiring.

Hence, from the original quest emerged the differential notions of metaphysics and epistemology. In order to understand the former, it was necessary to become skillful in the latter. As system after system developed to understand the mysteries of nature and its subsumption concerning the learning process, increasing emphasis was devoted to the methodological component of the respective system. Gradually, as methods became more reliable in terms of repetition of results, and valid in terms of applicability to observable phenomena, the original quest into substantive nature was recast into a methodological frame of reference in which nature could be understood only as a direct result of the method of inquiry employed. With this modified concept of his original task, man again sought answers with renewed vigor. For with this change came the possibility of assigning numbers to quantities. As this procedure became more sophisticated, ground rules for the legitimate inference of the nature of things from the manipulable methods employed became more formalized. One can see this transition from Goclenian syllogisms and the Organon to inductive propositions and the New Organon to the postulates of the Hullian hypothetico-deductive system. The knowledge-to-chaos ratio was to be understood in terms of the scientific methods using "descriptions," "predictions," and "control" as desiderata as well as operational values in, curiously enough, both the methodological and substantive components of the inquiry. I say "curiously" this has occurred, but I should have said as a logical extension, because as the original inquiry into the substantive what of things became recast into a question of the methodological how, the values of the original question become recast into the values required of the method. Consequently, at the present time in psychology, for a question to be "meaningful"—a term borrowed from the logical positivists—it must be cast into a testable hypothesis susceptible to the ministrations of the scientific technique used, and answerable only in this context. Skinner has pointed out that science carries with it its own ethic, and this has become painfully obvious in psychology. To illustrate with a familiar example, most graduate students select their research topics not on the basis of any substantive due of the problem, but on the basis of whether or not the problem will fit a chi square, analysis of variance, factor analysis, or other design. It is in this very point that I think that psychology as a whole has gone astray, and if I may extend a bit of doggerel—psychology not only first lost its soul, then its mind, then consciousness, but now has lost its purpose.

The nature of the learning process and the functionalistic ethic

Using these concepts of description, prediction, and control, experimentally derived and statistically treated data were fed into the scientific arcana. Usable knowledge about the nature of things proliferated and still does. In psychology (the most complex of any of the disciplines to which sophisticated methodological treatments are applied) important information about the nature of man accumulated. Explanations using empirically obtained descriptive data have been given as to the nature of man as a whole and the various subsystems of which he seems comprised. At the heart of all these views has been one central question: the nature of the learning process. For without this complex capacity to interact with and modify the environment as a separate self-reflexive, interdependent force system, man would not even be able to pose these tortuous problems to himself.

The nature of these explanations revolving about the learning process has been quite interesting in and of itself. These explanations have taken many forms. They have covered the philosophical gamut from animistic speculation which evolved into a deductive *a priori* system having religious overtones, with a hypostatized concept of cognition as a companion of conation and affection as seen in tripartite psychology (and its Freudian logical extension of reified dynamisms) to a concept of learning as a painstakingly slow concatenation of elements empirically seen to be agglutinized by bonds of association onto a *tabula rasa* having religious overtones with external concepts of stimulus and response, as seen in British empiricism and its latter day extension, behaviorism. To add flavor and provide further insight as to that which provided the power or drive for the system, postulations concerning perceptual reorganization arising from anisotropic field forces were added to the former centralist position, while abientadient constructs, arising from a misunderstood hedonistic dictum, unleashed by the presence of appropriate seriatim stimuli, were added to the latter peripheral position. By inserting the proper refinements—sometimes in the wrong places —innate releasing mechanisms have become confused with homeostatic balance, and secondary reinforcement with drive stimuli. But as these explanations became more sophisticated, and as methods by which they were derived became more sensitive, the large encompassing systems became smaller in scope and tighter in internal consistency. Thus, in modern psychology, models rather than theories have emerged with the above mentioned functionalism as an ethic. The preponderant conclusion of psychologists has been that it is better to know something, *i.e.*, something isolatable, manipulable, and predictable, about the microcosm of behaviors than to know little about the macrocosm of behavior. That this choice has been made is again, I think, an indication that the values of a methodology have supplanted those of a substantive oriented inquiry. But there is a further concomitant of the functionalist ethic. This has been the introduction in sampling theory of the opportunistic sample. This type of sample has come to be considered representative of a population from which to gather the data conducing to the empirically derived model being generated. Although institutionalized persons, aberrants, and conscripted college sophomores are seemingly—like the poor—ubiquitous, they nevertheless serve as a somewhat tenuous base for extrapolations concerning the nature of the learning process. Nowhere is this more true than in gerontology. Nor are investigators unaware of this, but beset by the Hobson's choice of this type of sample or no sample at all, the functionalistic ethic exacts its toll.

An inescapable conclusion

Consequently, from the probabilistically couched conclusions of the theoretical models which have emerged from albeit serious and dedicated research concerning the learning process, an inescapable conclusion can be reached to the question posed in the title of this paper, *i.e., What is the import of learning theory for gerontology?* That conclusion is simply that at the present time there is none. Despite the ingenious experiments of Gladis and Braun (1958), Melton (1940), and Gibson (1940) in transfer, the efforts of Botwinick (1959), Verzar-McDougall (1955, 1957) and others in motivation, the developmental contributions of Pressey (1957) and Kuhlen (1959), and the postulation of a model for the aging organism as refined by Welford (1958), the contention is still that there is currently no import for gerontology. This conclusion is not solely a syntactical tour de force based on the point that gerontology is still a substantive inquiry—albeit ill defined, problem-oriented and interdisciplinary—whereas learning theory is currently a methodological exercise comprised of a collection of isolated models based on different samples with a technique-generated and technique-dominated philosophy, which happens to be centered about the common element of a behaving organism. This difference alone would suffice to support the conclusion. But there is an additional point in which lies the heart of the whole gerontological movement, and which, consequently, is part of the psychologist's concern for the nature of the learning process. Unless this critical point is at least recognized, no learning theory, model, principle, hypothesis or investigation has any real meaning. McFarland put it succinctly and cogently when he asked, "Learning for what?" The question gives rise to a differentiation between a theory of aging and a theory of successful aging, but this is more appropriately treated in other chapters of Part I. But this question of why the status-less and societally purposeless aged person *should* learn is not a morbid question, nor one to sound the tocsin of doom for further investigation. Rather, it raises the central issue and problem of the study of the learning process as a function of aging at all phases of the life span. That issue is the one of motivation. To glean eclectically from conclusions of some of the many investigators doing work in the field, it becomes evident that even tentative statements concerning the learning process are qualified by reservations concerning the motivation, set, or other central attitudinal components of the aged subject. Such divergent investigators as Halstead (1963), Yarrow (1963); Axelrod (1963), Anderson (1963); Shamovian (1963), Hovland (1951); Welford (1958), Botwinick (1959); and Verzar-McDougall (1955, 1957), all allude to this factor. Jerome (1959) has epitomized it well in describing the motivation-learning-age syndrome as being as bound together in function as it is neglected in investigation. Kay (1959), in his recent reviews, comes to basically the same conclusion, although more euphemistically phrased and with a positive note of hope for the future. Consequently, the conclusion is reiterated to the question of import: at the present time, none.

Purity and substance

However, there seems to me a far more amenable and fruitful question—amenable in the very terms of the methodology which has so gripped psychology; amenable and fruitful in that the answer to it may provide the avenue by which the substantive values lost in the concern for methodological purity may be returned. That question is merely the reversal of the subject and predicate of the original title. What is the import of gerontology for learning theory? Here I feel is a challenge for the theoretician or model builder. The essential ingredients of a theory are:
1. Metricizable data descriptive of the phenomena.
2. Equivalence of data and constructs.
3. Internal consistency.
4. Wheels of inferential movement within the system.
5. External coherence and applicability.

These points, although simply phrased, are decidedly difficult to collate since they entail imposing an artificial static construction upon possibly an unequally dynamic process. Most learning theorists have paid little more than lip service to this static-dynamic dichotomy, possibly because of dualistic implications, but more probably because it does not suit the functionalistic ethic. There are some, however, who have recognized this and attempted to devise systems wherein the aging process itself is the end rather than a means to the end. These systems can generically be called organismic theories, and any future learning theories, to account for data of the aging organism, must, in my opinion, be of this genre. The sequacious cycle of the organism begins with a constellation of loosely confined forces and ends with a disruption and dispersion of this configuration. These are simply the terminal points of birth and death. But the entire range of the intervening behavior of this loose collection of forces called the organism is a legitimate interest and work area for the learning theorist. Any theory or model must of needs take into consideration the learning functions in all phases of this intervening process with attendant attention to fluctuations, whether they be quotidian, diurnal, seasonal, menarchical, or senile. A far simpler analogue is that of the moving averages of the Dow theory. Moreover, most theories and models have, unfortunately, been derived from data taken from cross-sectional slices of this process. This should not be construed as denigrating the heuristic value of these studies, nor as an argument for greater interest in longitudinal studies *per se*, although the research of Halstead (1947) and Bayley (1955) especially unearth some fascinating, if nonmodal, results—but should be considered as emphasizing not only the inadequacy of the findings, but the intrinsic fallacy of solely a methodological approach to answer a question which is essentially substantive in nature, *i.e.*, "Why should the aged organism learn?" Consequently, the lesson and thereby the import of gerontology for learning theory is that here, at last, the other end of the life continuum is being given the careful scrutiny and investigation that has been lavished heretofore on the initial periods of the life cycle. To flesh the total organismic skeleton, these data are crucial. It is of interest to note that many of those who have given the impetus to gerontology were in fact child developmentalists who, shall I say, have continued to develop. For it is with information provided from both terminal phases of the life cycle and the intervening stages, by both cross-sectional and longitudinal approaches, and with a substantively oriented system, that a consistent and coherent theory of learning may be erected. It is, then, with this redirected point of view that the future rapprochement of learning theory and gerontology lies. But it cannot be constructed without the nature of the process itself—a substantive value—being considered part of the framework which the theory must encompass.

In conclusion, let it be stated that while the import of learning theory for gerontology may be null, the import of gerontology for learning theory is infinite.

References

Anderson, J. E., (Ed.), *Psychological aspects of aging.* Washington D. C.: APA, 1956.

Anderson, J. E. Environment and meaningful activity. In R. H. Williams, C. Tibbitts, & W. Donohue (Eds.), *Processes of aging,* New York: Atherton Press, 1963.

Axelrod, S. Cognitive tasks in several modalities. In R. H. Williams, et al. (Eds.), *Processes of aging,* New York: Atherton Press, 1963.

Bayley, N. The maintenance of intellectual ability in gifted adults. *J. Geront.,* 1955, *10,* M-107.

Berelson, Bernard, Steiner. *Human behavior: An inventory of scientific findings.* New York: Harcourt Brace and World, 1964.

Birren, J. E. *The psychology of aging.* Englewood Cliffs, N. J.: Prentice Hall, 1964.

Botwinick, J. Drives, expectancies, and emotions. In Birren (Ed.), *Handbook of aging and the individual,* Chicago: University of Chicago Press, 1959.

Gibson, E. J. A systematic application of the concepts of generalization and differentiation to verbal learning. *Psychol. Rev.,* 1940, *47,* 196-229.

Gladis, M., & Braun, H. W. Age differences in transfer and retroaction as a function of intertask response similarity. *J. Exper. Psychol.,* 1958, *55,* 25-30.

Halstead, W. C. *Brain and Intelligence.* Chicago: University of Chicago Press, 1947.

Halstead, W. C. The Halstead index and differential aging. In R. H. Williams, et al. (Eds.), *Process of aging,* New York: Atherton Press, 1963.

Hilgard, E. R. *Theories of learning.* New York: Appleton-Century Crofts, 1956.

Hovland, C. I. Human learning and retention. In Stevens (Ed.), *Handbook of experimental psychology.* New York: John Wiley & Sons, 1951.

Jerome, E. A. Age and learning—experimental studies. In Birren (Ed.), *Handbook of aging and the individual.* Chicago: University of Chicago Press, 1959.

Kay, H. Theories of learning and aging. In Birren (Ed.), *Handbook of aging and the individual.* Chicago: University of Chicago Press, 1959.

Kuhlen, R. G. Aging and life adjustment. In Birren (Ed.), *Handbook of aging and the individual.* Chicago: University of Chicago Press, 1959.

Melton, A. W., & Irwin, J. The influence of degree of interpolated learning on retroactive inhibition and the overt transfer of specific responses. *Am. J. Psychol.,* 1940, *53,* 173-203.

Miller, G. A. *Mathematics and psychology.* New York: John Wiley & Sons, 1964.

Osgood, C. R. *Method and theory in experimental psychology.* New York: Oxford University Press, 1956.

Postman, L. (Ed.), *Psychology in the making.* New York: Alfred Knopf, 1962.

Pressey, S. L., & Kuhlen, R. G. *Psychological development through the life span.* New York: Harper and Brothers, 1957.

Shamovian, B. M., & Busse, E. W. Psychophysiological techniques in the study of aging. In R. H. Williams, et al. (Eds.), *Processes of aging.* New York: Atherton Press, 1963.

Stevens, S. S. *Handbook of experimental psychology.* New York: John Wiley & Sons, 1951.

Verzar-McDougall, E. J. Learning and memory tests in young and old rats. In *Old age in the modern world.* Edinburg: E. & J. Livingstone, 1955.

Verzar-McDougall, E. J. Studies in learning and memory in aging rats. *Gerontologia,* 1957, *1,* 65-85.

Welford, A. T. *Aging and human skill.* London: Oxford University Press, 1958.

Yarrow, M. R. Appraising environment. In R. H. Williams, et al. (Eds.), *Processes of aging.* New York: Atherton Press, 1963.

PART II
Biological Perspectives

AGING THEORY: Cellular and Extracellular Modalities

5

Harry Sobel

One of the fundamental questions in gerontology is whether non-reproducing cells have attributes of potential immortality or whether time alone imposes self-limitations. To put the question another way: do the causes of death lie within or outside the cells? Does cellular death occur independently of the environment or do environmental factors contribute significantly to it? These factors, of course, refer to all external conditions and influences which, with the exception of time, affect cells. They include invariants such as gravity, normal atmospheric pressure and composition, and thermodynamic considerations under physiologic normalcy. They also include variables arising from the gross environment to which the total organism is exposed (nutritional factors, environmental temperature, immunologically active and infectious agents, toxic factors, background radiation, and all interactions with physical and biological systems from outside the organism, etc.) and those of the microenvironment of the cells (adjacent cells, interstitial spaces and connective tissue, oxygen, nutrients, and products of other cells, including metabolic products, hormones, and antibodies). The distinction between time-induced changes which arise intrinsically and those which derive from the cellular environment is not simply a philosophical one, for the latter may be subject to some control. It should not be thought, however, that only one set of changes is implied to the exclusion of the other.

It is recognized that a knowledge of subcellular and molecular events brings one closer to an understanding of biological phenomena. This is almost a self-evident conclusion. Yet, in an attempt to obtain such information with regard to the aging processes, it seems that differences between young and old may not be very great at the molecular level. According to Dr. Nathan Shock (personal communication) the dramatic changes which one sees in the intact aging organism seem to become less and less impressive as the system is broken down into smaller and smaller components. The changes in the organized whole apparently exceed in magnitude the summation of those which occur at the subcellular level. This is a significant point since there is the danger that with the current surge of interest in molecular events insufficient attention will be given to aging of intact organs and organisms. The death of cells, however, must certainly have fundamental causes which reside at the subcellular and molecular level.

This review will cover some of the observations, working hypotheses and theories of events which occur within the cells and within the microenvironment of the cells with time. It may be argued that events taking place outside the cells might produce changes within them. This could be true in some cases.

A. Cellular events

It has been hypothesized that certain molecules which have a long half-life may become modified with time because of kinetic phenomena which may result in an increased number of hydrogen bonds and ester groups, etc., and because of the action of foreign molecules such as free radicals, metal ions, aldehydes, and oxygen. These produce cross-links between molecules so that they lose their enzymatic activity or other biochemical characteristics. If DNA were involved, mutations might result or there could be effects on the transmission of genetic information. The chief exponent of the concept of irreversible cross-linkages of protein is Bjorksten (1963) who has pointed out that human beings contain 20 parts per million of substances capable of cross-linking proteins or nucleic acids. They may reach the cell from without or may be generated by the cell itself. One characteristic of these cross-linked molecules, according to this theory, is that there is no way of getting rid of them; therefore, they may accumulate in the cell. Such accumulations have been reported in liver cells by Bjorksten and co-workers (1960) and in the protozoan Tokophrya by Rudzinska and Porter (1955).

The accumulation of lipofuscin—a lipoprotein material commonly designated "age pigment"—has been known for many years. In humans it accrues with advancing age in liver (Casselman, 1951), nerves (Hydén & Linström, 1950), and heart (Strehler, 1959). It also accumulates in primitive forms such as rotifera (Lansing, A. I., personal communication). It is unclear, however, whether this is the result of the passage of time alone or whether it is due to the experiences to which the organism has been exposed during this time.

The accumulation of metals has also been observed. Cells contain about eight metals which are either associated with enzyme action or with conformation of subcellular constituents. During the aging processes, body tissues accumulate and concentrate at least 20 other metals (Furst, 1964).

Age changes are being sought in enzymatic activity and, more directly, in the coding of information for protein synthesis (Medvedev, 1962). An interesting question has been raised with regard to alterations induced in cells because of environmental requirements. The need for "adaptation" may result in marked variation in protein synthesis. Cells may change their composition and activities in order to accommodate work demands placed upon them (Sobel & Cohen, 1958). When such changes have become programmed into a cell, the question arises as to whether or not some permanent residues result even after the need for such adaptation has passed. Some studies which have been carried out in this laboratory suggest that a "memory" of an event may remain long after the evocator has been removed (Sobel, Haberfelde, & Reeves, 1965). It is an obvious projection that "memories" of past events could influence future events.

The cells contain various *structural* elements which have been hypothesized to play some role in aging phenomena. Lysosomes contain a variety of lytic enzymes enclosed in a membrane. Substances which are ingested by the cell are attacked inside these "enzyme bags." The polymorphism of the lysosomes is due, in part at least, to the variety of substances and objects in various states of digestion. The enzymes may be used to control environment of the cell or they may be involved in metabolic events within the cell. However, when the membrane ruptures, as it may in cells suddenly deprived of oxygen or exposed to certain substances, the enzymes are released into the cytoplasm with disastrous results (de Duve, 1963). Furthermore, they may engulf large amounts of substances which they are not equipped to digest. The possible role of lysosomes in lipofuscin formation has been suggested (Samorajski et al., 1964).

The average cell contains several hundred mitochondria which carry out the function of generating and transforming energy. Although a great deal is known about the structure and function of the mitochondria (Lehninger, 1964) very little is known about the factors which regulate their number and activity. It is known that they are subject to environmental control. For example, in response to hypoxia, mitochondrial protein in the heart is increased (Sobel & Cohen, 1958). Only recently has evidence been obtained that mitochondria are self-duplicating (Gibor & Granick, 1964). The literature concerning the effect of age on the mitochondria of non-replaceable cells is confusing. Some authors have reported changes in number, form, and function. Others do not agree. Weinbach and Garbus (1956) suggest the possibility that in some tissues there is a shift in phosphate metabolism from oxidative to glycolytic exergonic reactions accompanying senescence.

The ribosomes are involved in protein synthesis following instructions from messenger RNA. The protein factories of the cells are known to be collections of ribosomes working together in orderly fashion: the polyribosomes. The metabolic events pertaining to these organelles is little understood, and it has been suggested that they may become "damaged" in some way in old non-reproducing cells with the result that protein synthesis in such cells may be affected. However, there is no evidence of this in the hearts of old dogs (Sobel, Thomas, & Masserman, 1964).

Membrane systems which surround the cytoplasm or organelles, such as the mitochondria, contain two membranes separated by a space, the membrane structure being composed of structural protein and phospholipid in a network arrangement. Protein molecules are attached to the membranes, and they contain systems that can move ions. The problem of renewal of cell membranes has not yet been resolved. It has often been suggested that membrane permeability may be affected with age. However, there is very little known about this matter.

B. Microenvironmental changes

The above are examples of some of the changes which are thought to take place within the cells with age. The role of extracellular factors in causing, regulating, or accelerating such changes is now receiving intense consideration. The domain of the microenvironment has been mentioned in the foregoing. The known age-changes in the microenvironment include arteriolocapillary fibrosis (Casarett, 1960) and a number of changes in the extracellular-extravascular milieu.

Virtually all cells are separated from their sources of oxygen and nutrients by a variety of barriers. These include the capillary endothelium, endothelial cell coatings, the glycocalyx, basement membrane, connective tissue cells and fibers, extracellular fluid—ground substance which consists of at least two phases and glycocalyx, and basement membrane around the cells themselves. The effect of age upon connective tissue components has been summarized (Sobel, 1965). Collagen may increase due to chronologic and pathologic accrual. There is a change in the physical-chemical characteristics of collagen with time which tends to decrease solubility and turnover. This is related to an increased number of cross-links (Verzár, 1963). Some connective tissues exhibit an increase in calcium and other minerals with age (Yu and Blumenthal, 1963). A number of changes occur in the ground substance which may be summarized as follows:

1. Reduction of ground substance space including total water and colloid mass.
2. Alteration in the compartmentalization of water so that there is a relative decrease in the "water-rich phase" and a relative increase in the "colloid-rich phase."
3. Decrease in the total mass of the acid glycosaminoglycans and their protein complexes.
4. Marked alteration in relative distribution of individual species of acid glycosaminoglycans. These changes seem to be in a direction which would decrease the water-holding capacity of the ground substance and modify colloidal charges.
5. Modifications in the physical characteristics of acid glycosaminoglycans-proteins suggesting decreased solubility and, perhaps, increasing metabolic inertia.
6. Changes in the glycoproteins which are not yet understood.
7. Increased thickness of the basement membranes.
8. Decrease in rate of transfer of fluid through the ground substance and return of fluid and contents through the lymphatic system.

It has been hypothesized that as a consequence of the increase in collagen and the loss of the ground substance the fibrillar density of connective tissue would increase (Sobel & Marmorston, 1956). This and other changes in the ground substance could affect the rate of delivery of oxygen, nutrients, hormones, and the removal of wastes. It was recently demonstrated that the rate of plasma clearance and the transvascular passage into the myocardium of I^{131}-albumin decreased with age in dogs (Sobel, Masserman, & Parsa, 1964). While this observation has no bearing on oxygenation and nutrition, it suggests that in the heart, at least, the rate of transvascular passage of plasma albumin and molecules of similar size could be reduced.

While the above must remain an hypothesis until supporting evidence is found, it can be concluded that the observed age-changes in the ground substance may affect various functions of the many which the connective tissues serve. These changes are probably cofactors in the genesis of age-associated disease processes. The question of whether or not they may also be involved in the age-changes within the cells and in the loss of irreplaceable cells is of great significance to gerontology.

The products of other cells reach the cells through the microenvironment and they include several factors which are thought to play some role in the aging process—the hormones, the production of which is often markedly age-dependent (Engle & Pincus, 1956), and autoantibodies. Autoantibodies may play a significant role in the loss or damage of irreplaceable cells and they are now the center of an active field of investigation (Walford, 1965).

C. Some implications

There are several aspects of what has been discussed which are of direct interest to the psychologist. It is commonly held that there is a redundancy in the oxygenation of tissue. That this is not the case is immediately apparent from the observation that a relatively small increase in energy expenditure results in an abrupt rise in lactate excess. According to Miller et al. (1964), under ordinary conditions the brain receives barely enough oxygen to maintain normal function. What would happen if oxygen content were reduced for any reason? McFarland (1963) has suggested that hypoxia and aging cause similar changes according to studies on (1) light sensitivity and dark adaptation, (2) critical flicker frequency, (3) effect of glare, (4) auditory sensitivity, and (5) mental functions. The causes could be related to reduced supply, delivery, diffusion, or utilization of oxygen. Aside from atherosclerotic and other vascular phenomena the question whether or not age-induced changes in connective tissue could account for these observations in the nervous tissues affected must be answered negatively since the brain does not appear to have a significant ground substance, although Van Harreveld and Crowell (1964) have recently found some evidence to support its existence. The effect of age on the perivascular glia is not known. The explanation of McFarland's correlation is awaited with great expectation.

It was stated before that cells may change their composition and activities in order to accommodate work demands placed upon them. It has been shown that proteins of the mammalian central nervous system, taken as a whole, have a turnover rate which, even though not as high, was not far below that of other metabolically highly active organs such as the liver (Waelsch, 1963). The protein metabolism of the nervous system resembles in many aspects the protein metabolism of any mammalian organ, although it has specific aspects as well. Although less is known about this matter with regard to nerve cells than with regard to some other cells, the evidence seems to suggest that challenge is desirable for cells in order to maintain their function, and that lack of stimulation is detrimental. The rapid onset of deleterious effects with sensory deprivation has been established. Proteins participate in the function of the central nervous system. These findings may have great significance with regard to the effects of retirement upon older individuals and with regard to the process of disengagement. Removal of daily challenges may accelerate or engender deleterious changes in the nervous system just as it does in the musculature.

References

Bjorksten, J. Aging, primary mechanism. *Gerontologia,* 1963, *8,* 179-192.

Bjorksten, J., Andrews, F., Bailey, J., & Trenk, B. Fundamentals of aging: immobilization of proteins in whole-body irradiated white rats. *J. Amer. Geriatr. Soc.,* 1960, *8,* 37-47.

Casarett, G. W. Acceleration of aging by ionizing radiations. In B. L. Strehler (Ed.), *The biology of aging.* Washington, D. C.: Amer. Inst. Biol. Sci. 147-152, 1960.

Casselman, W. G. B. The *in vitro* preparation and histochemical properties of substances resembling ceroid. *J. Exp. Med.* 1951, *94,* 549.

de Duve, C. The lysosome. *Sci. Amer.,* 1963, *208,* 64-72.

Engle, E. T., & Pincus, G. (Ed's.). *Hormones and the aging process.* New York: Academic Press, 1956.

Furst, A. Quoted in "Environmental variables in disease." *Science,* 1964, *146,* 954.

Gibor, A., & Granick, S. Plastids and mitochondria: inheritable systems. *Science,* 1964, *145,* 890-897.

Hydén, H., & Linström, B. Studies on yellow nerve cell pigment. *Disc. Faraday Soc.,* 1950, *9,* 436-441.

Lehninger, A. L. *The mitochondrion.* New York: Benjamin, Inc., 1964.

McFarland, R. A. Experimental evidence of the relationship between aging and oxygen want; in search of a theory of aging. *Ergonomics,* 1963, *6,* 339-366.

Medvedev, ZH. A. Aging at the molecular level and some speculations concerning maintaining the functioning of systems for replicating specific macromolecules. In N. W. Shock (Ed.), *Biological aspects of aging.* New York: Columbia University Press, 1962.

Miller, J. A., Jr., Zakhory, R., & Miller, F. S. Hypothermia, asphyxia, and cardiac glycogen in guinea pigs. *Science,* 1964, *144,* 1226-1227.

Rudzinska, M. A., & Porter, K. R. Observations on the fine structure of the macronucleus of Tokophrya infusionum. *J. Biophys. Biochem. Cytol.*, 1955, *1*, 421-428.
Samorajski, T., Keefe, J. R., & Ordy, J. M. Intracellular localization of lipofuscin age pigments in the nervous system. *J. Geront.*, 1964, *19*, 262-276.
Sobel, H. Aging of ground substance in connective tissue. *Advances in gerontological research*, 2. New York: Academic Press, 1965, in press.
Sobel, H., & Cohen, F. Proteins of the heart in experimental cardiac hypertrophy in the rat. *Proc. Soc. Exper. Biol. & Med.*, 1958, *99*, 656-658.
Sobel, H., Haberfelde, G., & Reeves, A. Reversibility of endocrine changes produced in guinea pigs by exposure to cold. *Amer. J. Physiol.*, 1965, *208:1*, 115-117.
Sobel, H., & Marmorston, J. The possible role of the gel-fibre ratio of connective tissue in the aging process. *J. Geront.*, 1956, *11*, 2-7.
Sobel, H., Masserman, R., & Parsa, K. Effect of age on the transvascular passage of I^{131}-labeled albumin in hearts of dogs. *J. Geront.*, 1964, *19*, 501-504.
Sobel, H., Thomas, H., & Masserman, R. Myocardial proteins in dogs of various ages. *Proc. Soc. Exp. Biol. & Med.*, 1964, *116*, 918-921.
Strehler, B. L., Mark, D. D., Mildvan, A. S., & Gee, M. W. Rate and magnitude of age pigment accumulation in the human myocardium. *J. Geront.*, 1959, *14*, 430-439.
Van Harreveld, A., & Crowell, J. Extracellular space in central nervous tissue. *Fed. Proc.*, 1964, *23*, 304.
Verzár, F. The aging of collagen. *Sci. Amer.*, 1963, *208*, 104-114.
Waelsch, H. Protein metabolism of the nervous system. *Swiss Med. J.*, 1963, *93*, 1-13.
Walford, R. L. Immunology and aging. This volume.
Weinbach, E. C., & Garbus, J. Age and oxidative phosphorylation in rat liver and brain. *Nature*, 1956, *178*, 1225-1226.
Yu, S. Y., & Blumenthal, H. T. The calcification of elastic fibers. I. Biochemical studies. *J. Geront.*, 1963, *18*, 119-126.

THE SOMATIC MUTATION THEORY OF AGING   6

Howard J. Curtis

The basic theory and its problems

Through the years there have been a great many theories to explain why it is that a person grows old. Each of them has made an important contribution to our understanding of the subject and each has been superseded by other theories as experimental evidence has shown them to be inadequate.

Recently a theory was advanced, known as the somatic mutation theory of aging, which has great intellectual appeal as a theory, but until very recently there has been practically no evidence either favoring or opposing it. The general idea of the theory is quite simple. Spontaneous mutations are postulated to occur in the somatic cells of the body, and since they are, in general, irreversible, their numbers will tend to accumulate with age. Since every mutation tends to curtail some cellular function, the cells will gradually become inefficient or die as the mutations accumulate. When this process takes place in a sufficiently large percentage of the cells of the body, senescence gradually develops.

Research carried out at Brookhaven National Laboratory under the auspices of the U.S. Atomic Energy Commission.

The problem with this theory has been that it has not been possible to test it, because no methods are available for measuring mutations in somatic cells. Further, if one takes the known rates for gene mutations in mammals and computes the numbers of mutations expected in the somatic cells on the same basis, one arrives at a figure of an order of magnitude too low to account for the phenomenon of senescence. Nevertheless, the idea is so attractive that several alternate suggestions have been made to get around this difficulty. Failla (1958) computes how large the mutation rate will have to be for somatic cells to explain aging, and does not feel the figure is unreasonably high. On the other hand, Szilard (1959) assumes that only chromosome aberrations constitute large enough faults to affect aging, and that in diploid organisms such as man, there are some numbers of inborn faults in the chromosomes of an individual. Each fault is on only one of a pair of homologous chromosomes. As chromosome aberrations increase with age, the probability of having some vital cellular function eliminated grows steadily larger. This theory involves a number of assumptions which would be very difficult to prove.

An experimental approach

Several years ago we embarked on an experimental approach to the problem. In plants the somatic cells differentiate to form germ cells. One can score the numbers of visible chromosome aberrations in the somatic cells under a variety of conditions and relate these to the true mutations, scored in the next generation. When this was done it was found that in every case the chromosome aberrations observed in the somatic cells were proportional to the true mutations scored in the next generation (Caldecott, 1961). If this holds true for plants it should hold true also for animals. Consequently, we developed a method for scoring chromosome aberrations in somatic cells, and have been taking this as an index of the numbers of mutations present in this cell population.

The technique consists simply in inducing regeneration and, thus, cell division in the liver by destroying part of the organ by an injection of CCl_4. When regeneration is at its height, the animals (mice) are sacrificed. Liver squashes are made, and the cells observed in anaphase or early telophase are scored as either normal or abnormal. Almost all abnormalities consisted of bridges or fragments. One typical abnormal anaphase is shown in Figure 6-1.

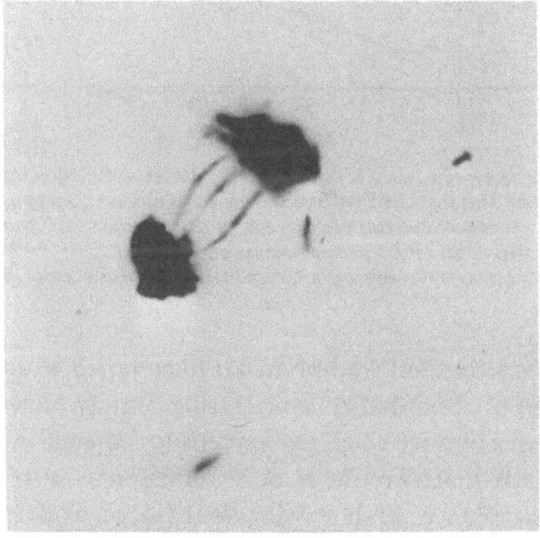

Figure 6-1 Photomicrograph of an abnormal dividing liver cell exhibiting 3 bridges and 3 fragments (from Curtis, 1963).

Using this technique, a number of different situations have been investigated. It was found that aberrations increase steadily with age (Stevenson & Curtis, 1961. *See* Figure 6-2).

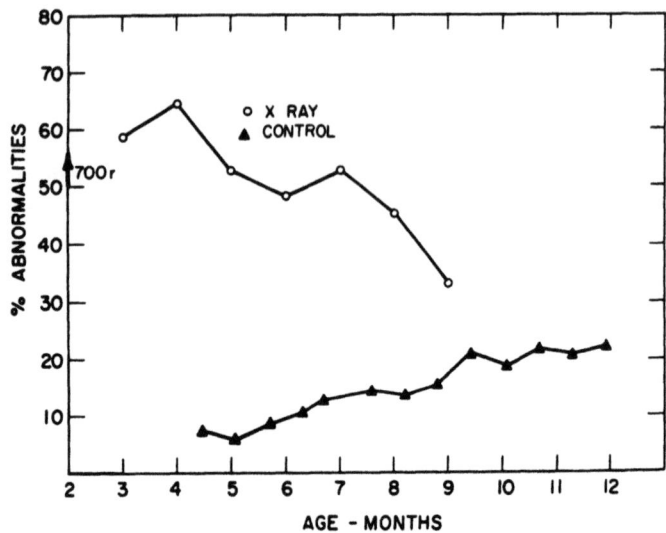

Figure 6-2 Chromosome aberrations in liver cells as a function of age in normal mice and in mice that had received a large but non-lethal dose of x rays. The curves show the steady increase in the number of chromosome aberrations (mutations) with natural aging and the dramatic increase and slow return to normal following a dose of radiation (from Stevenson & Curtis, 1961).

This is a very constant finding in all mice tested so far, and the numbers of abnormal cells reach surprisingly high values in some strains of mice when the animals become old.

It is known that chronic X or γ radiation is only about 25 percent as effective in shortening the life span as is an acute dose of the same size. It is found that the same is true for the production of chromosome aberrations (Figure 6-3). It is also

SOMATIC MUTATION THEORY

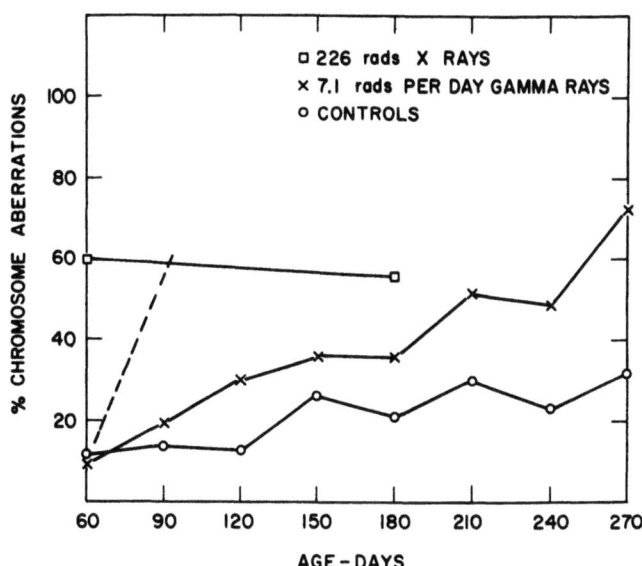

Figure 6-3 Chromosome aberrations in liver cells of 1) mice subjected to chronic gamma irradiation, 2) mice given a single dose of x rays, and 3) normal mice. The dashed line shows the rate of build up which would be expected for the mice of group 1 if chronic were as effective in producing chromosome aberrations as is acute irradiation. Since the experimental curve has a much smaller slope it is concluded that chromosome healing takes place following small doses of radiation (from Curtis & Crowley, 1963).

known that chronic neutron irradiation is just as effective in shortening the life span as is acute neutron irradiation. Again, the same has been found for the production of chromosome aberrations (Curtis, Tilley & Crowley, 1964. *See* Figure 6-4).

It is well known that some inbred strains of mice are quite short-lived. It was found that chromosome aberrations increase quite rapidly in one short-lived strain, and quite slowly in a long-lived strain (Crowley & Curtis, 1963. *See* Figure 6-5).

Thus in every situation tried so far there is a quantitative relation between the degree of life shortening and the de-

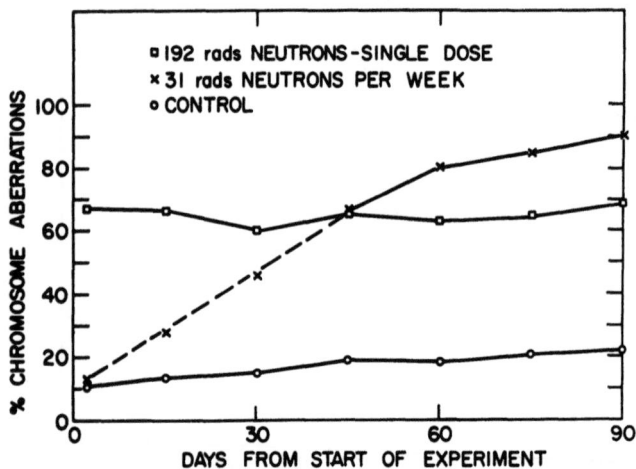

Figure 6-4 Chromosome aberrations as a function of time for 1) mice subjected to chronic neutron irradiation, 2) mice subjected to acute neutron irradiation, and 3) control mice. The dashed line gives the rate of build up of aberrations in group 1 mice which would be expected if chronic were as effective in producing aberrations as is acute irradiation. It will be seen that the experimental points fall closely on this line. The data show there is no chromosomal healing following even small doses of neutrons (from Curtis, Tilley & Crowley, 1964).

velopment of somatic mutations in liver cells. This would seem to indicate strong support for the somatic mutation theory of aging and, indeed, it does. But at the same time it raises a number of questions which require modification of the simple theory.

Further developments in the somatic mutation theory

One of the problems is the long delay between the production of the mutation and its manifestation in terms of an increased death rate. A dose of radiation given to a young mouse may

SOMATIC MUTATION THEORY

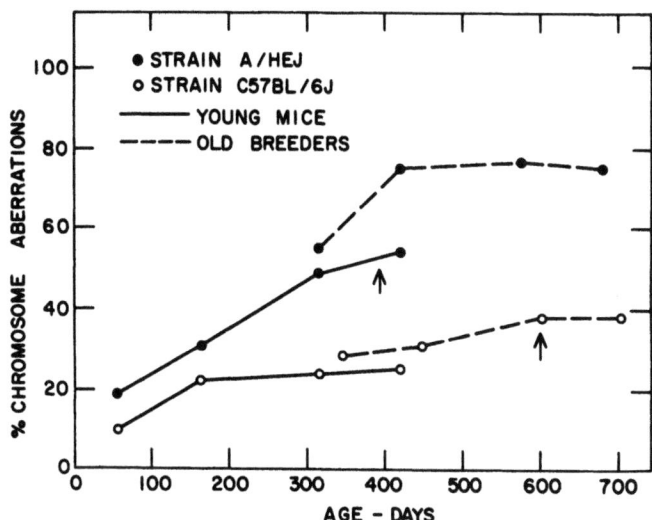

Figure 6-5 Chromosome aberrations in liver cells of normal mice of two different inbred strains as a function of age. The median life span of each strain is indicated by the arrows. In each case the solid lines represent animals 8 weeks of age at the start of the experiment, and the broken lines old breeding animals about 1 year old at the start of the experiment (from Crowley & Curtis, 1963).

produce serious chromosomal damage in the vast majority of the cells of the body, yet the mouse appears unaffected and may not show signs of aging for many months, although he will do so before his control. At first glance one would expect that if aging is due to mutations it should be accelerated as soon as the mutations are produced.

Although there are several possible explanations to account for this apparent contradiction, it is best explained in terms of the modern concepts of the function of DNA and RNA in cell function. It is believed that for each cell function there is a single DNA molecule in the genome. This molecule synthesizes the corresponding protein (enzyme) which controls the function in question. If the DNA molecule in this chain is damaged, the entire process will eventually cease, and the cell will become inefficient or die. But if a large number of the RNA or protein molecules in the chain are damaged, the cell can recover. Even if the DNA is damaged it may take a long time for all the protein and RNA to be used up. This sequence can explain how the destruction of only a very few molecules, perhaps even only one, can cause the destruction of a cell, and why there may be a long delay between the insult and the manifestation.

This forms an attractive picture of the reasons why even heavily irradiated mice that have their chromosomes severely damaged can lead apparently normal lives for a long time, but will eventually become senescent and die. Yet, direct proof of this explanation is lacking. Most work with RNA seems to indicate quite a short turnover time. However, this is from work with systems such as bacteria that have a very short cell cycle. Work with highly differentiated cells indicates a very much longer turnover time, and there is some indication that the turnover time may be comparable to the division cycle of the cell, which for some highly differentiated cells is comparable to the life time of the animal. There is, however, no doubt that some cells, such as the mammalian red blood cell, can live for long times with no nucleus at all. It thus seems reasonable to accept this picture until it is proven incorrect or until a better one is available. This question has been extensively reviewed elsewhere (Curtis, 1963).

Another problem, already mentioned, is that of chromosome healing. One can imagine that as a group of cells undergoes cell division the cells containing mutations are gradually eliminated by cell selection and the cell line is thereby kept pure. This unquestionably happens, and in the mammal several organs fall in this category. The bone marrow, for example, contains virtually no aberrant cells even in very old animals, but immediately following a dose of radiation one sees many aberrant cells which are soon eliminated. Thus this organ appears to be immortal from this point of view, and as far as is known it would continue to function indefinitely provided it is well supported. Other organs such as the intestinal epithelium and the epidermis are essentially similar. However, the exception to this general statement is the cancer problem, since these organs are the ones which develop cancer, and cancer is a late sequela of radiation damage. It appears that among the mutations which occur either spontaneously or radiation-induced, a very few produce cells which have a selective advantage for survival and do not respond to the constraints on cell division which govern the size of normal organs. This is, of course, the mutation theory of carcinogenesis which fits the present observations very well.

The non-dividing cells of the body will accumulate mutations which they are unable to get rid of, and, depending on the particular function which the mutation affects, the cell may become inefficient, may die, or may have its function virtually unaffected. The effect is, on the average, bad and probably accounts for senescence in these organs. But since there is no cell division, there is no possibility of cancer development.

This indicates that these two types of mammalian cells play quite different roles in the process of senescence. The dividing cells are able to rejuvenate themselves by cell division but may develop cancer, partly because of an unfavorable mutation they are unable to eliminate. The non-dividing cells gradually accumulate mutations and become senescent. Thus,

I feel that the old question, as to whether individual cells age or not, has been answered—in the mammal they do.

The work with chronic gamma irradiation (Figure 6-3) shows that the individual cell can recover at least to some extent from the effects of radiation. This means the individual chromosomes have at least a limited ability to repair themselves following minor damage. More recent experiments (Curtis, Crowley, & Tilley, 1964), have shown that this repair process can proceed at the chromosomal level for many months provided the cells are not required to undergo cell division. Apparently these somatic cells are constantly developing flaws in the chromosome structure which are then repaired provided the flaw is not too large or serious. The large ones remain unrepaired and are responsible for senescence. Thus, the molecular stability of the chromosome is primarily responsible for senescence, and the repair processes which return them to their stable conditions are as important as the inherent stability of the structure. We know that some animals have quite a stable chromosome structure and are long-lived, while some have quite an unstable structure and are short-lived. We do not know all the factors responsible for molecular stability, but know there must be ways of influencing it. This gives us great hope for the future.

The somatic mutation theory explains a number of well known phenomena very nicely. In Drosophila, mice, and men it has been found that the offspring of old mothers have more defects and live a shorter life than those of young mothers; for example, mongolism, which is known to be caused by a genetic defect, is many times more prevalent in children of mothers over age 40 than in children of younger mothers. If the age of the father has any effect in this regard, it is quite small. On the basis of the present evidence it would seem that, since the oocytes in the female stay in the ovary for years without undergoing division, they have no opportunity to throw off mutations. Mutations accumulate in these cells as time goes on, some of which survive meiosis to endow the offspring with mutations. In males, spermatogenesis is in progress continually, thus mutations tend to be eliminated by cell selection.

The results of these experiments do not allow a computation of the mutation rates in somatic cells, but the rate in the normal animal must be very high. If the rate were anything like as high in the germ cells, the species could hardly survive a generation. The gametic cells must either have very much more stable chromosomes than somatic cells, or else meiosis and the various factors associated with fertilization constitute a very rigid screening process which allows only genetically perfect cells to get through. The bulk of the evidence today indicates the latter to be the case, so genetic methods measure only a very small fraction of the true mutation rate in germ cells. One of the chief objections to the somatic mutation theory was that the mutation rate was thought to be too low. The present experiments eliminate this objection.

Whatever the meaning of death for the individual, a moment's reflection will convince one that death is necessary for the good of the species. The fact that the somatic mutation rate is so much higher than the observed gametic mutation rate may be nature's method of insuring the death of the individual and the survival of the species.

Summary

Aging is a slow deterioration of the individual which manifests itself by an increased susceptibility of the organism to all kinds of disease. Of the many theories of aging, the somatic mutation theory seems to offer the best explanation of the known facts of aging. Recently, a method has been developed for estimating the mutations present in the somatic cells of mice. In all situations tested so far, a quantitative relation has been found between aging and the development of mutations in somatic cells. This mutation rate is very much higher than that measured for gametic cells and would seem to be ample to account for the phenomenon of aging.

References

Caldecott, R. S. Seedling height, oxygen availability storage and temperature: Their relation to radiation induced genetic injury in barley. In *Effects of ionizing radiations on seeds.* Vienna: IAEA, 1961.

Crowley, C., & Curtis, H. J. The development of somatic mutations in mice with age. *Proc. Nat. Acad. Sci. U.S.*, 1963, *49*, 626-628.

Curtis, H. J. The late effects of radiation. *Proc. Am. Phil. Soc.*, 1963 (a), *107*, 5-10.

Curtis, H. J. Biological mechanisms underlying the aging process. *Science,* 1963 (b), *141*, 686-694.

Curtis, H. J., & Crowley, C. Chromosome aberrations in liver cells in relation to the somatic mutation theory of aging. *Radiation Res.*, 1963, *19*, 337-344.

Curtis, H. J., Crowley, C., & Tilley, J. The elimination of chromosome aberrations in liver cells by cell division. *Radiation Res.*, 1964, *22*, 730-734.

Curtis, H. J., Tilley, J., & Crowley, C. The cellular differences between acute and chronic neutron and gamma ray irradiated mice. In *Biological effects of neutron and proton irradiations.* Vol. II, pp. 143-155, International Atomic Energy Agency, Vienna, 1964.

Failla, G. Mutation theory of aging. *Ann. N.Y. Acad. Sci.*, 1958, *71*, 1124-1138.

Stevenson, K. G., & Curtis, H. J. Chromosomal aberrations in irradiated and nitrogen mustard treated mice. *Radiation Res.*, 1961, *15*, 774-784.

Szilard, L. On the nature of the aging process. *Proc. Natl. Acad. Sci. U.S.*, 1959, *45*, 30-42.

IMMUNOLOGY AND AGING 7

Roy L. Walford

Changes in immunological status with advancing age

A considerable body of evidence exists suggesting a heightened and/or dysfunctional immunological status with advancing age in vertebrates (Walford, 1962, 1964). There is an impressive increase in serum gamma globulin with age in all animals thus far studied, including humans, rats, gerbils, and bulls (Riegle & Neller, 1964). The percentage change in gamma globulin is relatively greater than the changes in most other biochemical parameters of aging, particularly if one omits values for very young animals. A relative increase in the ratio of splenic weight to body weight occurs with age in most animals. This increase is accompanied by a decrease in mitotic activity of splenic cells, but the decrease is considerably *less* than corresponding decreases in mitotic activity in other organs that contain dividing cell populations.

Evidence such as the above for a relative augmentation of immunological activity with age might be explained on the basis of increased random processes, heightened antibody production in response to specific infective or other agents, or by

---

Supported by NIH Grant No. HD-00534.

developing low grade generalized autoimmune response. There is no evidence for an increased output with age of antibodies directed against infective agents or foreign antigens. Iso-agglutinin titers against blood group agglutinogens actually decline with age. However, those antibodies which are generally characteristic of various autoimmune states do, in fact, increase substantially in incidence with advancing years among subjects *who are clinically normal.*

Antinuclear antibodies, the rheumatoid factor, antibodies against gastric parietal cells, antithyroid antibodies, and anti-insulin antibodies all increase in frequency with age in subjects who display no clinical evidence of the various corresponding diseases (Walford, 1964). The relatively heightened immunologic activity in the latter portion of life may thus be directed against components of the body's own tissues. Amyloidosis, which is present to a variable degree in some old specimens of all vertebrate species, may be an autoimmune disease. Even in those species thought to show only a low incidence of amyloidosis with age, specific and sensitive fluorescent chemical staining of old tissues reveals in fact quite a high incidence of the malady. In old men, for example, examination of tissue stained with the fluorescent dye thioflavine-T has yielded evidence of amyloidosis in over 80 percent of cases.

On the basis of observations such as these, it has been proposed that aging is a consequence of increasing immunogenetic diversification of the dividing cell population with age. These diverse cells lose the ability to recognize "self," and a low grade, prolonged histoincompatibility reaction sets in, analogous to a chronic autoimmune state, and manifested as aging. Figure 7-1 will serve to explain this concept. Assume that one is dealing with a mouse heterozygous at the H-1 histocompatibility locus. One allele at this locus possesses the

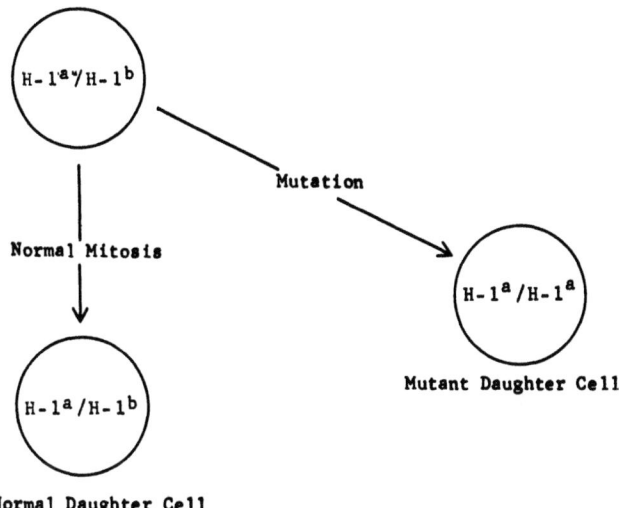

Figure 7-1 Schema to illustrate mutation at a single histocompatibility locus in a heterozygote. Mutant daughter cell *not* recognized by the body as foreign, but itself recognizing the body as foreign.

H-1^a gene and the other allele the H-1^b gene. If one of these alleles undergoes a mutation to the other, then the daughter cell would be, for example, H-1^a/H-1^a. This cell would not be recognized by the body as foreign and, therefore, would not be eliminated via classical transplantation disease mechanisms. However, the cell would on its part recognize the body as foreign and might react against it. The process is analogous to the graft-versus-host reaction, which in a broad sense may be considered a form of autoimmune disease. There is reasonable evidence on the basis of statistical genetics that such mutations might occur in the body to a very significant degree (Failla, 1958).

Implications of the immunogenetic hypothesis

Several implications of the immunogenetic hypothesis may be pointed out. A mutation at a histocompatibility or other type of immunologic locus would not necessarily be intrinsically detrimental to the cell. An H-1^a/H-1^a cell is a perfectly normal cell. Its residence in an H-1^a/H-1^b body is, on the other hand, an abnormal situation. The cell, however, would not be rejected by the body, or be under any competitive disadvantage. Antigenic mutation is thus not conceived as necessarily leading to a "loss of vitality" of the mutating cell. Such cells, arising in the body, would not necessarily be eliminated with further cell division, as would be the case if the mutation were intrinsically deleterious to the cell's chemical machinery.

One possible advantage of an immunologic theory of aging is its susceptibility to direct experimentation. Along this line, we have injected newborn H-1^a mice intracardially with adult immunologically competent H-1^b spleen cells. Under these circumstances the newborn animals are rendered tolerant of the adult foreign cells. The foreign cells might be expected to reside and multiply in the newborn animals, and perhaps react against these animals. This is clearly a "model" experiment, an attempt to mimic artificially the process hypothesized in Figure 7-1. In such an experiment we have noted that the test animals show no change in weight or mortality until ten to twelve months of age. At this age, the test animals begin dying at an accelerated rate compared to control animals injected with H-1^a cells. The peak incidence of lymphoma in the injected test animals is also shifted to the left. A curve of "accelerated aging" is thus produced which resembles the life shortening induced by X-irradiation. The result may be considered as tentative support for an immune theory of aging.

A second type of "model" experiment consists of joining two animals in parabiosis which differ one from the other at only relatively weak histocompatibility loci. For this experiment we have used hamsters of the Byzanson strain. In such joined animals one observes, after about 20 to 80 weeks in parabiosis, the development of severe amyloidosis in chronologically young animals. This disease pattern is perhaps a noteworthy finding because the hamster, while quite resistant to production of amyloidosis by the usual experimental means (injection of sodium caseinate or other agents), nevertheless displays an exceedingly high incidence of senile amyloidosis. Amyloidosis produced by parabiosis is morphologically identical to that produced in the hamster by senility. Thus the chief "disease of aging" in the hamster was produced by the second "model" experiment. It has also been possible to cause amyloidosis in chronologically nonsenile hamsters by means of X-irradiation. Irradiation with 425 r at three months of age led to a mean survival time of 14 months, as compared to 22 months in nonirradiated controls. The life-shortening produced in the hamsters by this degree of irradiation was accompanied by an apparent shift in incidence of amyloidosis to a younger age. This result may tentatively be taken as further evidence, in addition to the $H-1^b \rightarrow H-1^a$ mouse experiment, for some relation between life-shortening produced by immune mechanisms and by X-irradiation.

Two possible mechanisms for aging on an immune basis can be suggested. The first is mitotic inhibition. There is ample evidence for slowing of the mitotic rate with age. The number of mitoses per unit area in various organs clearly decreases with age (Korenchevsky, 1961). Lesher et al. (1961) documented an increased generation time of the proliferative cells of the small intestine with age. The mitotic rate with age of the lens fiber cells declines markedly (Verzar,

1963). Now Cole et al. (1962) presented evidence that many of the pathologic manifestations of transplantation disease might be explicable in terms of mitotic arrest. The second possible mechanism for aging on an immune basis concerns the idea that antibodies may act as cross-linking or denaturing agents. This point has been discussed elsewhere (Walford, 1964). Suffice it to say that some antigens following combination with antibody seem more resistant to enzymic degradation than uncombined antigens.

A further step that must be taken with regard to age-altering immunologic experiments, in order to support or negate the theory, is to show that such alterations are accompanied by changes in biochemical or other parameters which are characteristic of normal aging. These should particularly include changes in collagen and liver enzymes. Work along these lines is in progress.

References

Cole, L. J., Nowell, P. C., & Davis, W. E., Jr. Suppression of wound healing by graft-vs.-host reaction in mice with transplantation disease. *Fed. Proc.*, 1962, *21*, 37.

Failla, G. The aging process and cancerogenesis. *Annals New York Acad. Sci.*, 1958, *71*, 1124-1140.

Korenchevsky, V. *Physiological and pathological aging.* New York: Hafner Publishing Company, 1961, p. 54.

Lesher, S., Fry, R. J. M., & Kohn, H. I. Age and the generation time of the mouse duodenal epithelial cell. *Exper. Cell. Res.*, 1961, *24*, 334-343.

Riegle, G. D., & Neller, J. E. Effect of age upon blood cellular and protein components. *Fed. Proc.*, 1964, *23*, 212.

Verzar, F. *Lectures on experimental gerontology.* Chicago: Charles C. Thomas, 1963, p. 88.

Walford, R. L. Auto-immunity and aging. *J. Gerontol.*, 1962, *17*, 281-285.

Walford, R. L. Further considerations towards an immunologic theory of aging. *Exper. Gerontol.* 1964, *1*, 67-76.

CHROMOSOMAL CHANGES AND AGING

8

Lissy F. Jarvik

Other chapters in this volume have placed varying emphasis upon intracellular and extracellular events in creating those changes which are characteristically associated with human aging. As our understanding advances, we shall be able to assign appropriate weights to the many processes of aging and to theories stressing the loss of long-lived, irreplaceable cells (*e.g.*, neurons); molecluar changes in deoxyribonucleic acid (*e.g.*, cross-linkage); accumulation of somatic mutations, immunological changes (*e.g.*, autoimmunity), hormonal changes (especially as a result of stress), or the "rate of living" with particular emphasis upon the relation of brain weight and body size (Pearl, 1924; McCay et al., 1939; Shock, 1951; Sacher, 1959; Selye & Prioreschi, 1960; Walford, 1962; Comfort, 1963; Curtis, 1964).

Mitotic errors and aging

The hypothesis to be discussed here is based upon changes in the nature of cell populations constituting the human organism and upon the age-related effects which may be attributable to the presence of cells which ought to be replaced but are not. In man, the white blood cells—specifically, the lymphocytes—

appear to constitute a system in which cells with gross chromosomal anomalies not only survive, but initiate clonal progeny which propagate the aberration. With a view to Dr. Walford's theory based on age-related immunologic disharmony, (see Chapter 7), it seems only proper that lymphocytes—cells which are known to participate in antibody production— should be the cells used for cytogenetic research in gerontology. Actually, the selection of lymphocytes was made without this knowledge and dictated largely by considerations of practicality.

Lymphocytes are to human cytogenetics what the fruitfly (Drosophila) is to classical genetics. Both are easily procured and maintained, multiply readily under laboratory conditions, and have a short generation time. To grow lymphocytes in suspension culture requires a minimum of specialized equipment, and the cells undergo division within three days. Lymphocytes are readily obtained by taking a sample of venous blood, just as is done for any number of routine blood tests. Most subjects will consent to this familiar and innocuous procedure. Attempts to remove any other type of tissue from healthy human beings encounter formidable obstacles, whether the technique involves introducing a needle into the bone marrow, removing a piece of tonsil or taking a skin biopsy.

Figure 8-1 depicts the chromosomes of one such cell grown in tissue culture. In the lower right-hand corner is a picture of the chromosomes as seen under the microscope. From an enlargement of this photograph the individual chromosomes are cut out and arranged according to size and position of the centromere (central constriction). The resulting karyotype, occupying the main portion of the picture, enables us to identify the individual chromosomes and to detect which, if any, is missing and which, if any, is present in excess. The normal human diploid complement, illustrated in Figures 8-1 and 8-2, consists of 46 chromosomes—44 autosomes and two sex chromosomes. The cell shown in Figure 8-1 was derived from a male (XY), that in Figure 8-2 from a female (XX).

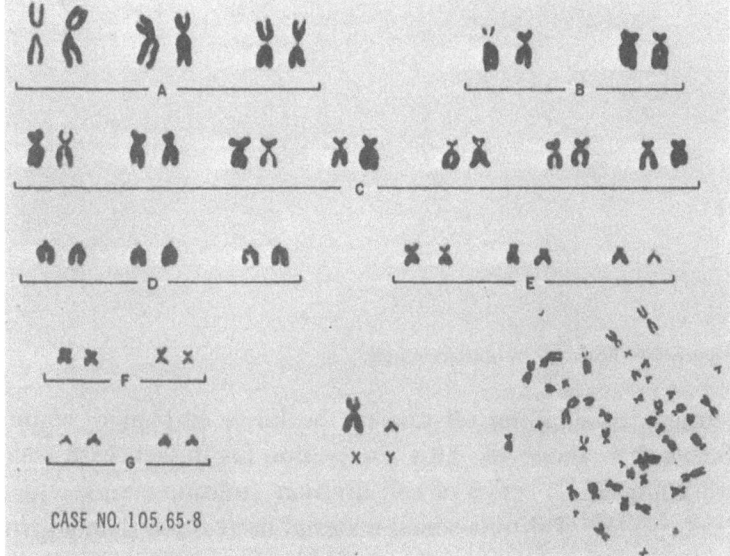

Figure 8-1 Karotype of normal male.

Contrary to past teaching that there could be no deviation from the diploid number in a normal individual, it has now become clear that there are exceptions to this rule, as there are to many rules. Although the vast majority of cells derived from normal persons contain 46 chromosomes, not all of them do—some cells having less and some cells more than 46. The distribution of chromosome counts resembles the normal

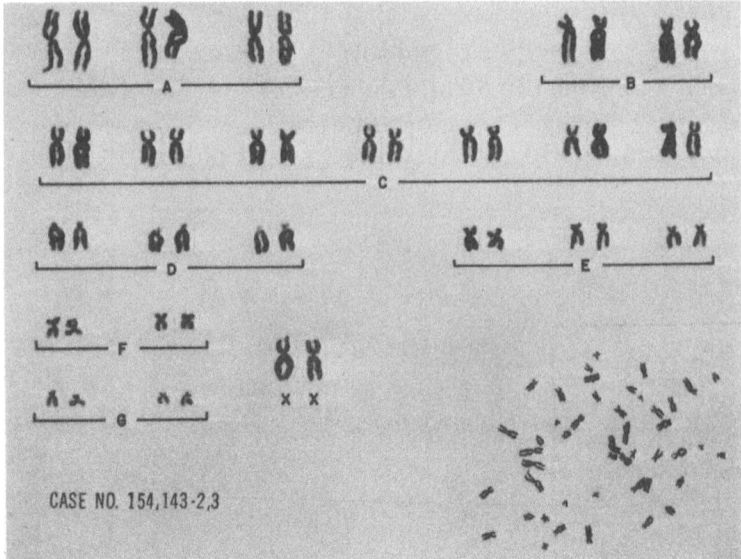

Figure 8-2 Karotype of normal female.

curve with a tailing off toward the lower end—more counts below than above 46. This observation fits in well with what we know about errors of cell division (mitotic errors): they lead to a loss of chromosomal material more often than a gain.

Assuming that errors of cell division occur throughout the life of the organism, an accumulation of such errors could well influence the processes of aging. Since mitotic errors (by definition) can occur only in those cells that undergo division, and since the long-lived, irreplaceable cells divide only rarely, if at all, we are concerned chiefly with short-lived, rapidly dividing cellular systems.

Interestingly enough, mitotic errors do not show a random occurrence. Since 1959, investigators throughout the world have examined hundreds of thousands of human cells and have observed abnormalities for only half a dozen of the 23 chromosome pairs. It is the same chromosomal anomaly that is found again and again. Hence, mitotic errors compatible with clonal reproduction are probably selective and not random. If repetition of the *same* error has a high statistical probability, then the hypothesis of mitotic error can account for some of the similarities observed among aged persons and particularly among one-egg twin partners. Such similarity applies not only to the visible physical signs of aging, but also to intellectual performance and to total length of life.

Heredity, intelligence, and survival

The inference that hereditary factors may be among the determinants of total length of life stems from a number of statistical investigations conducted since the beginning of this century. Despite the application of varying statistical techniques to different population groups, all of these studies demonstrated a positive relationship between parental age and filial life span (summarized by Kallmann & Jarvik, 1959, and Cohen, 1964), while the results of a longitudinal study of aging twins revealed that even after the age of 80 the average differences in survival of two-egg twin partners still exceeded those of one-egg twins (Jarvik et al., 1960). In the latter long-term follow-up study, over 2,000 twins aged 60 years or older were collected throughout the State of New York between 1946 and 1949. Analyses of life-span differences, carried out during the succeeding 12 years, showed that mean intra-pair differences declined with increasing age, as expected from the asymptotic nature of the survival curve. Nonetheless, greater mean intra-pair life span differences were observed in dizygotic than in monozygotic pairs of both sexes, and for all age groups above age 60. Such consistency would be expected by chance less than once in 1,000 times ($1/2^{10}$), and led to the conclusion that gene-specific elements significantly affect man's natural life span.

With respect to intellectual functions, the results of twin studies have shown that even after the age of 60 years one-egg twins are significantly more alike than two-egg twins in performance on a number of intelligence tests (Jarvik et al., 1957). Of particular interest in this connection is the case history of one pair which had outstanding differences in living conditions and which was followed for over 15 years. These twins were separated at the age of 18, one of them marrying a local farmer, the other entering missionary school. The married twin raised a large family (six children) and in her entire life was never further away from her birthplace than ten miles. The co-twin, soon after she finished a course in Bible school, was sent as a missionary to the Orient where she spent much of her adult life. Consequently, there were marked differences not only in climate and diet, but also in the sociocultural milieu. Yet, when the twins were reunited after the retirement of the missionary, following 47 years of separation, they were so similar in physical appearance that people had difficulty in distinguishing them. This similarity obtained despite a weight difference of 28 lbs. in favor of the widow. The degree of their intellectual resemblance is illustrated by the test scores which were obtained when the twins were 88 years old (Table 8-1). Differences in theoretical and esthetic values reflected their diverse experiences, while otherwise the resemblances were far more impressive than the differences. The missionary died at the age of 92 following a fractured hip due to a fall, while the widow survived for another 28 months until the age of 94.

Table 8-1 Test scores of twins at age 88.

Subtests from Wechsler-Bellevue Scale

	Vocabulary (Stanford-Binet)	Digits Forward	Digits Backward	Similarities	Digit Symbol Substitution	Block Designs
Missionary	34	5	4	17	23.5	9
Housewife	34	4	4	14	23.0	9

Study of Values (Allport-Vernon-Lindzey)

	Theoretical	Economic	Aesthetic	Social	Political	Religious
Missionary	31	36	43	47	33	50
Housewife	41	39	29	45	41	45

There is some evidence indicating an association between certain types of intellectual changes during senescence and remaining survival time. Thus, in our longitudinal study of senescent twins a higher level of performance was originally achieved on the tests by subjects who survived than by those who died during the follow-up period (Jarvik et al., 1957). Further, stability of scores on certain tests of cognitive function (particularly Vocabulary, Digit Symbol Substitution, and Similarities Subscales of the Wechsler-Bellevue) appeared to be associated with five-year survival, *i.e.*, five years after the last test session (Jarvik, 1962).

Indications of a connection between survival and intellectual performance have also been reported by a number of investigators who studied single-born subjects (Kleemeier, 1961; Sanderson & Inglis, 1961; Berkowitz & Green, 1963; Birren, 1964). Since variations in both intellectual performance and natural life span are known to reflect genotypic differences, the existence of common underlying mechanisms may be postulated.

Possible mechanisms of aging

Thus, we may entertain the hypothesis that one such common denominator consists of the gradual accumulation of aberrant cells (*i.e.*, clonal propagation of mitotic errors), with the eventual attainment of a level at which homeostatic mechanisms can no longer cope with primary and secondary metabolic dysfunctions, a level incompatible with survival (Jarvik, 1963). Experimental support for this hypothesis stems from the observation of a statistically significant increase in the proportion of aneuploid cells (*i.e.*, cells containing other than the normal diploid number of chromosomes) with increasing age (Jacobs et al., 1961; 1964). Aneuploidy was noted in less than five percent of cells at ages below 25 years and over 15 percent for ages above 75 years.

The changes occurring in rapidly dividing cells would be expected to affect not only the specific intra- and intercellular environment, but also to exert secondary effects upon long-lived cells. Long-lived cells, then, would be available as targets for mutagenic agents, and, in addition, would be subject to the metabolic disturbances resulting from the greater aneuploidy of the short-lived cells.

There remains an obvious question. How can such marked change in chromosomal constitution, occurring with so high a frequency in somatic cells, be reconciled with the much-vaunted constancy of the germ plasm? Upon careful consideration, it seems that the chromosomal constitution of germ cells —or gametes—is actually far from constant. The incidence of mongolism (an extra chromosome #21—Group G), for example, is approximately 1 in 600. The frequency of Klinefelter's syndrome (an extra X chromosome) is approximately 1 in 500.

There are numerous other less common chromosomal anomalies in man. Yet, in aggregate, the frequency is at least 5 per 1,000. That means 5 per 1,000 *viable offspring.* How much more often the offspring are not viable, leading to abortion or miscarriage, and how much more often defective gametes are unable to produce a zygote, is unknown. It is likely, however, that the total frequency is much higher than the frequency of observed anomalies in viable fetuses. Thus, the germ plasm is not constant after all, and its relative constancy is maintained largely by the process of natural selection. Yet errors that are lethal in gametes, *i.e.,* at the unicellular level, may be merely detrimental in somatic cells, *i.e.,* at the multicellular level. *In the complex organism, then, aging may have replaced natural selection.*

As in a civilized society, less efficient and even defective cells are protected and aided by other members of the community to the best of their ability. When the proportion of deviant cells, increasing with the passage of time, reaches the point where interference with homeostatic mechanisms prevents the organism from coping with its responsibilities, the result is disease and death. It is reasonable to suppose that, with increasing age, changes take place both in the long-lived and in the short-lived cells; the exact nature and relative importance of these changes awaits evaluation.

Among the questions raised by the preceding discussion are the following:

1. Do cells from rapidly aging individuals show greater aneuploidy than cells from relatively healthy senescent persons? Certainly, patients with progeria or presenile psychoses

would be valuable for comparison as would be the well-preserved centenarians found in certain parts of the world.

2. How do the life spans of persons that have chromosomal abnormalities compare with those of karyotypically normal persons? Of special interest here would be abnormalities like Klinefelter's. Do these individuals follow the male or the female patterns of disease incidence and differential mortality?

3. Does the direct relation between increasing aneuploidy and increasing age hold for cells other than lymphocytes?

4. What is the correlation, if any, between chromosomal changes and psychological changes? Particularly useful for this purpose would be the comparison of twin partners with similar and with different rates of aging. It is tempting to take chromosomal aberrations, which amount to errors in the genetic code, and to relate them directly to deficiencies in nervous system functioning. Ribonucleic acid (RNA), for example, has been implicated in such a fashion, but caution is needed in the light of recent experimental findings.

In conclusion, we may suppose that the time sequences involved in aging are programmed into the individual at birth, like the predictable time sequences of other changes associated with the growth cycle. Sexual maturity, for example, almost always occurs in the early teens and is rarely delayed by more than a few years. Similarly, individuals in their seventh or eighth decades who maintain youthful appearances or functions are exceedingly rare. On the other hand, unlike the loss of deciduous teeth at a certain age or the appearance of menarche at another, aging phenomena, occurring after the reproductive period, cannot be regarded as adaptive processes determined by natural selection. To account for the heritable aspects of aging and longevity one may postulate the operation of pleiotropism, *i.e.*, that genes influencing aging exert their major effects earlier in life and on some other trait (*e.g.*, intellectual functioning) and only secondarily influence aging and longevity. Further, it is possible that the psychological changes in the senium are associated with aneuploid changes in the brain, particularly in the glial elements which, in contrast to neuronal elements, are known to have a high rate of mitosis.

References

Berkowitz, B., & Green, R. F. Changes in intellect with age: I. Longitudinal study of Wechsler-Bellevue scores. *J. Genet. Psychol.*, 1963, *103*, 3-21.

Birren, J. E. Neural basis of personal adjustment in aging. In P. F. Hansen (Ed.), *Age with a Future,* Copenhagen: Munksgaard, 1964.

Cohen, B. H. Family patterns of mortality and life span. *Quart. Rev. Biol.*, 1964, *39*, 130-181.

Comfort, A. Mutation, autoimmunity, and aging. *Lancet*, 1963, 2, 138-140.

Curtis, H. J. The biology of aging. *Brookhaven Lecture Series* (Brookhaven National Laboratory), No. 34, March 18, 1964.

Jacobs, P. A., Brunton, M., & Court Brown, W. M. Cytogenetic studies in leucocytes on the general population: subjects of ages 65 years and more. *Ann. Hum. Genet.*, 1964, *27*, 353-365, London.

Jacobs, P. A., Court Brown, W. M., & Doll, D. R. Distribution of human chromosome counts in relation to age. *Nature*, 1961, *191*, 1178-1180.

Jarvik, L. F. Biological differences in intellectual functioning. *Vita hum.*, 1962, *5*, 195-203.

Jarvik, L. F. Senescence and chromosomal changes. *Lancet*, 1963, *1*, 114-115.

Jarvik, L. F., Falek, A., Kallmann, F. J., & Lorge, I. Survival trends in a senescent twin population. *Am. J. Hum. Genet.*, 1960, *12*, 170-179.

Jarvik, L. F., Kallmann, F. J., Falek, A., & Klaber, M. M. Changing intellectual functions in senescent twins. *Acta Genet.,* 1957, *7,* 421-430.

Jarvik, L. F., Kallmann, F. J., Lorge, I., & Falek, A. Longitudinal study of intellectual changes in senescent twins. In C. Tibbitts and W. Donahue (Eds.), *Social and Psychological Aspects of Aging,* New York: Columbia University Press, 1962.

Kallman, F. J., & Jarvik, L. F. Individual differences in constitution and genetic background. In J. E. Birren (Ed.) *Handbook of Aging and the Individual: Psychological and Biological Aspects* Chicago: University of Chicago Press, 1959.

Kleemeier, R. W. Intellectual changes in the senium or death and the I.Q. Presidential Address, Division on Maturity and Old Age, American Psychological Association, New York, Sept., 1961.

McCay, G. M., Magnard, L. A., Sperling, H., & Barnes, L. L. Retarded growth, life span, ultimate body size and age changes in the albino rat after feeding diets restricted in calories. *J. Nutrition,* 1939, *18,* 1-13.

Pearl, R. *Studies in Human Biology.* Williams & Wilkins, Baltimore, 1924.

Sacher, G. A. Relation of lifespan to brain weight and body weight in mammals. In G. E. W. Wolstenholme and C. M. O'Connor (Eds.), *The Lifespan of Animals.* London: Ciba Foundation Colloquia on Aging. J. & A. Churchill, 1959.

Sanderson, R. E., & Inglis, J. Learning and mortality in elderly psychiatric patients. *J. Geront.,* 1961, *16,* 375-376.

Selye, H., & Prioreschi, P. Stress theory of aging. In N. W. Shock (Ed.), Aging: *Some Social and Biological Aspects,* American Association for the Advancement of Science, Washington, D.C., 1960.

Shock, N. W. Gerontology. *Ann. Rev. Psychol.,* 1951, *2,* 353-370.

Walford, R. L. Auto-immunity and aging. *J. Geront.,* 1962, *17,* 281-285.

ON LONGEVITY REGARDED AS AN ORGANIZED BEHAVIOR:
The Role of Brain Structure

9

George A. Sacher

Introduction

Most experimental biologists implicitly accept the proposition that biological aging can be explained by characterizing the changes that occur with time in the molecular composition of the organism. Since that proposition is a corollary of one of the basic postulates of molecular biology, the well-nigh universal assent accorded it by the present generation of experimental biologists is no cause for surprise. It is surprising, however, that contemporary psychologists accept the same viewpoint, despite the fact that in doing so they say in effect that the basic discoveries about aging will be made by the biophysicists and biochemists, and that the essentially psychological modes of inquiry are limited to the subordinate role, socially useful though it may be, of describing the deterioration of mental abilities with age.

A system of thought achieves such dominance only because it is productive and has not been successfully assailed. Knowing this, I nevertheless propose to show that the paradigm[1] of molecular biology, despite its great power in a number of

---

[1] In the sense of Kuhn (1962).
Work supported by the United States Atomic Energy Commission.

other fields, is inadequate and insufficient as a research strategy for biological as well as psychological gerontology.

Evidence will be presented that the rate of aging in mammalian species is dependent on the properties of the species phenotype as an integrated control system. These order or system properties are not presently reducible to molecular description in any meaningful sense, nor is there a prospect that such a reduction will soon be possible. The ways in which these system properties are operated on by the processes of evolutionary phylogeny, genetic selection and ontogenetic conditioning will be considered briefly.

This view is not a theory of aging but rather an alternative approach to the investigation of longevity and aging that is compatible with quite different or even opposed hypotheses about the mechanisms of development and aging in organisms regarded as active, self-organizing systems. Some possible behavioral hypotheses are considered.

It is perhaps desirable that one point not at issue be stated explicitly. Everything said here is based on an unqualified acceptance of the proposition that the properties of living systems are governed by the laws of quantum mechanics without any residue of vitalism or emergent supraphysical laws. More specifically, it is here accepted that all the changes observed in the course of psychological and somatic aging have a physical, *i.e.*, molecular basis.

Uncertainty as a parameter of living systems and its relation to the nervous system

For some years I have been working on a stochastic theory of mortality and aging which states that a major part of the total variance of the time-mortality or dosage-mortality curves manifested by animal populations is essentially random, in that it arises from an inherent uncertainty in the performances of living systems (Sacher, 1956; Sacher & Trucco, 1962). This uncertainty can be attributed in part to random errors in physiological control systems, and particularly to errors in such performances of the nervous system as discrimination, decision, memory and the neurohumoral control of effectors.

One implication of the theory is that the *stability* of a species phenotype,[2] defined as a duration and estimated by reciprocal mortality rates and mean or maximum longevity, depends in a quantitative, measurable way on the capability of the species for control and adaptation of its physiological performances. Direct test of this prediction would involve measuring the cybernetic capabilities of a number of species of mammals and relating these to their longevity characteristics. While this approach will one day be feasible, such a comparative cybernetics does not exist at present.

In default of a direct approach, an indirect approach was adopted. The mediating hypothesis was introduced that among *homologous*[3] species the weight of the brain is a valid

---

[2] For succinctness the term *species* is here used to denote not only taxonomic species but also lower groups such as subspecies or local races. In some cases a whole genus can be considered a species in regard to the relation of longevity to brain and body size. The *phenotype* is the typical or average behavioral and somatic expression of the species genome in an appropriate environment. An important requirement is that the species be a stable population with a high level of physiological efficiency and vigor maintained by strong selective pressures originating from the physical environment and from interspecies and intraspecies competition. Laboratory and domestic animal populations do not satisfy this requirement, nor does the living human race. This places restrictions on the use of data from such populations for the study of structure-function relationships.

[3] Two species are homologous in the degree that their organs and sub-organs (such as brain regions) are homologous, *i.e.*, have a similar structure, common embryological origin and a monophyletic evolutionary origin. The members of the Mammalia are completely homologous *inter se* but mammals and birds are not perfectly homologous.

measure of the brain's summated performance capability, when brain weight and performance are measured on healthy typical representatives of adapted populations.

Combining these two hypotheses yields the compound hypothesis that there is significant partial regression of species life span on adult brain weight, independent of body weight. It should be noted that the relationship will fail to occur if either of the two hypotheses is false.

This hypothesis was tested by examining intercorrelations of maximum life span, brain weight and body weight. There is a great store of such data in the zoological literature, so that there are now more than 100 species of mammals for which these 3 variables are tabulated. The hypothesis that life span has a positive partial correlation with brain weight is confirmed at a high confidence level (Sacher, 1959). The relation is found to be a power function, like the allometric relationships of somatic dimensions (Huxley, 1932) or Stevens's magnitude scale for prothetic continua (1961). The equation is:

$$[\text{life span}] \propto [\text{brain weight}]^{2/3} / [\text{body weight}]^{2/9}$$

This relation accounts for maximum life spans of terrestrial eutherian species with a maximum error of a factor of 2. It holds for mammals as different in body size as shrews and whales, or as different in level of cephalization as rodents and primates. The human life span is accurately estimated. The multiple correlation of life span with brain and body weight is 0.84. This is considerably less than perfect predictability, but a large part of the residual variance is assignable to error in the data, so that the correlation is in fact near the theoretical maximum. In the smaller group of species for which all three variables are accurately measured there is as yet no instance where the observed life span departs markedly from the predicted.

Implication for research strategy

The discovery of a functional relation between length of life and brain weight gives strong support to the two hypotheses that:

1. Species longevity depends on the cybernetic capability of the phenotype; and
2. the weight of a mammalian brain is an estimator of its cybernetic capability, equally valid for man and mouse.

Each of these verifications opens up a field of investigation. The first provides a sufficient justification for the inception of a program of behavior-oriented research on aging and its modification by genetic and ontogenetic means. Verification of the second hypothesis shows that a quantitative relation can be established between brain structure and performance if the performance can be scored quantitatively for a sufficiently wide range of species.

Evidence for a relationship between brain size and mental ability has been presented previously by B. Rensch and colleagues, who examined problem-solving abilities, memory and abstraction in species or subspecies that differ in brain weight. An association of learning ability and memory with brain weight was observed for iguana lizards (Rensch & Adrian-Hinsberg, 1963), domestic chickens (Altevogt, 1951), and trout (Saxena, 1960).

We now have presumptive evidence that the longevity and aging characteristics of mammalian species are based on supramolecular order properties that are not reducible to the thermodynamic properties of individual molecules. Consider the following concrete example. Man has a lifespan 3 times greater than that of the white-tailed deer or the mountain lion. The 3 species have about the same body weight, identical sets of homologous organs and tissues, about the same number and kind of proteins and nucleic acids and about the same metabolic average rate (Brody, 1945). Yet the human organism is capable of doing 3 times more metabolic work in its lifetime. Certainly all plausible hypotheses should be examined to discover the physical basis of this great difference in utilization of the soma by different species. There is reason to examine the hypothesis that the macromolecules of *Homo sapiens* are thermodynamically superior to those of *Odocoileus virginianus* or *Felis concolor,* just as there is reason to examine the redundancy hypothesis which says that the longer-lived species has more replicates of everything (Johnson, 1963). Nevertheless, it must be said that there is at present no experimental evidence for either of these hypotheses. On the other hand, the human brain is about 7- to 10-fold larger than the brains of these other 2 species, almost exactly the factor predicted by the regression relation. Nor should we forget that the brain is the only organ or structure thus far known to correlate with longevity.

Resistance to this conclusion is deep-seated in some quarters, and is not easily overcome by appeal to the accumulated evidence. Some biologists accept the empirical finding, but believe that it will eventually be given a more acceptable alternative interpretation in which brain size reflects some other, as yet undiscovered, nonbehavioral property. However, this view requires the additional assumption of a pleiotropic action of the genes for brain size.

A behavior-oriented approach to longevity and aging

I now wish to sketch some essential problems in the behavioral approach to aging. First, the prevailing physicalistic emphasis on aging as something that happens to an organism as a passive object is countered by an emphasis on those active performances or behaviors of the organism that maintain the dynamically stable living state. Longevity can be considered to be a measure of the quality of that overall performance in somewhat the sense that a person's intelligence test score or school performance measures his cognitive mental ability. Now that longevity has been shown to be a metrical quantity in homeothermic species, we can look for other measures of lifetime performance, or for alternative ways of scoring. The lifetime total of energy expenditure, number of cell divisions or information transcriptions, voluntary muscular activity, or number of consummatory acts are possibilities to examine. The part-scores, the unitary abilities that contribute to these complex performances, need to be identified, for it is these that genetic selection and ontogenetic training act upon.

Second, any gain on the behavioral side must be matched by like efforts on the structural side. Total brain weight is, like longevity, a composite metrical score. It can be regarded as a single-factor measure of the size of the whole information processing system, and the next step will be to identify the anatomical subsystems. Biometric analysis of the interrelations among brain parts is a rapidly developing research area, as may be seen in the excellent quantitative comparative studies that have appeared recently (Luetgemeier, 1962; Pilleri, 1960; Portmann, 1962; Stephan & Andy, 1964). The relation of brain weight to performance is only intelligible if brain weight is closely related to number of neurons. Fortunately, a close allometric relation of neuron size to brain size has been demonstrated (Bok & Kip, 1939; Tower, 1954).

It is important to determine whether the size of the brain also measures the informational capacity of the various non-neural control and adaptive systems. One important centralized information system is the immune system. This can be described as a molecular brain with high capability for discrimination and memory and a kind of learning and extinction behavior. Does its capacity in different species stand in some determinate ratio to brain size? Insofar as long life requires low susceptibility to infectious as well as degenerative disease, there should be a correlation, but no way of measuring it has yet been found.

Third, there are a number of questions about the phylogeny of longevity and brain size. One conclusion that can be drawn is that in consequence of natural selection the brain of every adapted species, whether it be large or small, is as efficient as it can be for its size. But why is it that size, and not larger or smaller? The same question arises in regard to longevity. This suggests the existence of a nested sequence of problems in optimum design. The formulation of this approach would illuminate the processes of macroevolution, especially in relation to the anagenetic increase in brain size that has occurred in vertebrate evolution (Rensch, 1959).

The behavioral approach is concerned less with the sequence of changes that take place in manifest aging than it is with how the rate and course of aging depend on the organism's genetic endowment of neural and sensorimotor structures and on the programming of this system during maturation. Thus, another direction of investigation is to determine whether nervous systems can be trained to improved physiological control in anything like the degree that they can be educated to more subtle cognitive relations with the external world.

The cognitive role of the nervous system, however, should not be allowed to overshadow its affective and conative role. That great gerontologist, William Shakespeare (King Lear, Act IV, Scene 1), put forth a profound and amazingly modern theory of aging in the lines:

> ". . . . . . O world!
> But that thy strange mutations make us hate thee,
> Life would not yield to age."

Out of this complex matrix of thoughts let us consider just the idea that the assaults of the environment lead to a progressive alienation from life. It would be out of place here to propose a restatement of this thought in psychophysiological terms, but it will be agreed, I am sure, that such translation could be achieved by several more or less plausible mechanisms. The length of time required for this highly conjectural process to reach the state of physiological acedia could reasonably depend on the size of certain brain structures. Unfortunately for this hypothesis, the brain structures that would seem to be most immediately implicated, such as the hypothalamus, reticular formation and hippocampus, are phylogenetically conservative brain regions and do not share the great allometric increase of the neopallial structures.

The point to remember is that we know in general terms how the brain functions as a life-sustaining organ, and yet we do not have any good measure of *how well* it functions, nor do we know what limits its function, or how it fails.

Conclusion

The purpose of this chapter has been to show how the demonstrated role of the brain in longevity opens up possibilities for investigating the ways in which an organism's ability to sense and control its own states governs the rate at which irreversible molecular degradations take place in its tissues. The longevity of the macromolecules of the organism is a function of its *milieu intérieur*. A synthesis of the molecular and molar approaches to gerontology can be achieved through the recognition that in mammalian species the most significant parameter of that *milieu* is the precision with which it is controlled by the neurohumoral integrative mechanisms of the species.

This synthesis, however, will not easily come about within the framework of a discipline which limits itself to the investigation of aging at the level of material causation. The prevalence of this paradigm in all domains of gerontology, psychological as well as biological, can be objectively verified by counting the frequency of occurrence in the Methods section of published papers of forms of words such as: ". . . 24 rats aged 60 days were . . . and 24 rats aged 600 days were . . ."

From the standpoint of the behavioral sciences, aging studies should be placed in the context of a discipline that as yet has no name. It could appropriately be called behavioral gerontology, with the understanding that its subject matter is not only the aging of behavior but also the dependence of physical senescence on the finite accuracy, adaptability and response rate of physiological behavior mechanisms. Somewhat more inclusively, this discipline can be defined as *the science of life in the aspect of its finitude.*

In this reference frame, aging and death are aspects of the

temporal finitude of the organism that are to be examined in terms of their dependence on the finite precision of the moment-to-moment behavioral acts of perception, decision and execution with respect to its internal environment. The goal of a remission of the aging process is approached through the hitherto unused insight that the more information the organism can acquire and utilize to control its own states the longer it will live.

The explicit recognition of human finitude is a necessary condition for the attainment of the goals of gerontological research. Although it is superficially a pessimistic view, in actuality it is hopeful in a sense that a dogmatically molecular outlook cannot be. There is, after all, overwhelming evidence that the limitations of the living state have been progressively transcended through the organic evolution of behavior and through the social evolution of human culture. This justifies the effort to determine whether present limits on human life cannot be transcended in some degree through the proper use of the organs of psychophysiological behavior, with the knowledge that these organs are by their nature accessible and educable.

References

Altevogt, R. Vergleichend-psychologische Untersuchungen an Hühnerrassen stark verschiedener Körpergrösse. *Zeitsch. f. Tierpsychol.*, 1951, *8*, 75-109.

Bok, S. T., & Kip, M. J. van Erp Taalman. The size of the body and the size and the number of the nerve cells in the cerebral cortex. *Acta Neerl. Morphol.*, 1939, *3*, 1-22.

Brody, S. *Bioenergetics and growth*. New York: Reinhold Publishing Corp., 1945.

Huxley, J. *Problems of relative growth*. New York: The Dial Press, 1932.

Johnson, H. Redundancy and biological aging. *Science*, 1963, *141*, 910-912.
Kuhn, T. S. *The structure of scientific revolutions*. Chicago: The University of Chicago Press, 1962.
Luetgemeier, F. Histologische Sonderheiten der Gehirne von Chiropteren verschiedener Körpergrösse und verschiedener Orientierungsweise. *Z. Morph. Ökol. Tiere*, 1962, *50*, 687-725.
Pilleri, A. Beitrag zur vergleichenden Morphologie des Nagetiergehirnes. *Acta Anatomica*: 1959, *Suppl. 38*, 1-124, Beiträge 1-3. 1960, *Suppl. 40*, 1-88, Beiträge 4-6.
Portmann, A. Cerebralisation und Ontogenese. In K. F. Bauer (Ed.) *Medizinische Grundlagenforschung*, vol. 4, pp. 1-62. Stuttgart: Georg Thieme Verlag, 1962.
Rensch, B. *Evolution above the species level*. New York: Columbia University Press, 1959.
Rensch, B. & Adrian-Hinsberg, C. Die visuelle Lernkapazität von Leguanen. *Zeitsch. f. Tierpsychol.*, 1963, *20*, 34-42.
Sacher, G. A. On the statistical nature of mortality, with especial reference to chronic radiation mortality. *Radiology*, 1956, *67*, 250-257.
Sacher, G. A. Relation of lifespan to brain weight and body weight in mammals. In Wolstenholme and Connor (Eds.) *Ciba Foundation Symposium on the lifespan of animals*, London: Churchill, 1959, pp. 115-133.
Sacher, G. A. & Trucco, E. The stochastic theory of mortality. *Annals New York Acad. Sci.*, 1962, *96*, 985-1007.
Saxena, A. Lernkapazität, Gedächtnis und Transpositionsvermögen bei Forellen. *Zoologische Jahrbücher*, 1960, *69*, 63-94.
Stephan, H. & Andy, O. J. Quantitative comparisons of brain structures from insectivores to primates. *Am. Zoologist*, 1964, *4*, 59-73.
Stevens, S. S. The psychophysics of sensory function. In W. Rosenblith (Ed.) *Sensory communication*. New York: John Wiley and Sons, 1961.
Tower, D. B. Structural and functional organization of mammalian cerebral cortex: The correlation of nervous density with brain size. *J. comp. Neurol.*, 1954, *101*, 19-52.

INDEX

Acid glycosaminoglycans, 63
Activation, intentional outside, 34
Activity, as adaptive technique, 41
Activity Theory, 34
Adaptive techniques, 40-42
   "Angry men", 41
   "Armored" group, 41
   "Mature" men, 25-26, 42
   "Rocking Chair," 25, 40-41
   "Self-haters", 41
Adjustment, interpersonal theory of, 37-43
Age pigment (lipofuscin), 59
Aging, definitions, 3
Albumin, in cells, 63-64
Amyloidosis, 82, 85
Anderson, J. E., 4-5
Aneuploidy, 94, 95, 96, 97
Antibody production, 81-82
Anxiety, and self-esteem, 37
Arteriolocapillary fibrosis, 62
Atherosclerosis, 65
Autoantibodies, 64

Basic trust, concept of, 9
Behavioral approach, to longevity, 105-107

Behavioral gerontology, 108
Behaviorism, 48
Bible, reading of, 27-28
Blood group agglutionogens, 82
Brain *see also* Nervous system
   aneuploid changes in, 97
   immune system and brain size, 106
   and longevity, 99-109
   neurons, 105-106
   oxygen supply, 64-65
   weight, in homologous species, 101-104

Cancer, 77
Capacity to delay gratification, concept of, 9
Carcinogenesis, mutation theory of, 77
Cells:
   acid glycosaminoglycans, 63
   albumin, 63-64
   arteriolocapillary fibrosis, 62
   atherosclerotic phenomena, 65
   autoantibodies, 64
   cellular events, 58-62
   chromosomes *see* Chromosomes
   collagen, 62, 63
   connective tissue, 62

DNA, 59, 75-76
enzymes, 60
fluid transfer, 63
ground substance, changes in, 62-64
hormones, 63, 64
hypoxia, 64-65
I^{131}-albumin, 63
lactate excess, 64
life span, 57
lipofuscin, 59, 60
liver, 59, 71-75
lysosomes, 60
lytic enzymes, 60
membrane systems, 61-62
metals, accumulation of, 59
microenvironmental changes, 62-64
mitochondria, 61
oxygen supply to brain, 64-65
oxygenation of tissue, 64
perivascular glia, 65
polyribosomes, 61
protein cross-linkages, 59
protein molecules, 61-62
protein synthesis in, 60
proteins of nervous system, 65
RNA, 61, 75-76, 96
radiation, effects of, 72-73
ribosomes, 61
somatic mutation theory, 69-80
structural elements, 60
water, 63
Chromosomes:
aneuploidy, 94, 95, 96, 97
aberrations, 70-75, 87-88, 96
changes in, 87-97
germ plasm, 94-95
healing, 76-77
heredity, 91-94
intellectual functions, 92-93
Klinefelter's syndrome, 95, 96
mechanisms of aging, 94-97
mitotic errors, 87-91
natural selection, 95
number in humans, 89-90
pleiotropism, 96-97
RNA, 96
survival and intellectual performance, 93-94
twins, 91-93, 96
Collagen, 62, 63
Connective tissue, 62
Cumming, E., 16

DNA, 59, 75-76
Death, perception of, 16
Depression, 42
Developmental events, 30-32
Developmental tasks, 31
Developmental theories, 4-6
Differentiation concept, 13
Disabilities, sensory, 40
Disengagement:
  as adaptive technique, 40-41
  and adult development, 15, 17, 19-35
  as aspect of normal aging, 16
  concepts and data, 19-26
  and protein metabolism, 65-66
Dow theory, 52
Drosophila, 78, 88

Ego energy, 22, 24, 26
Empiricism, British, 48
Employment, and self-esteem, 39-40
Engagement, as developmental event, 30-32
  see also Disengagement
Engrossment and perspective concepts, 7-13
Enzymes, 60
Erikson, Erik H., 37

Fromm, Erich, 37

INDEX

Functionalistic ethic, and learning theory, 47-49

Gamma globulin, changes in, 81
Gamma irradiation, 72, 73, 78
Generations, hostility between, 38-39
Genes, mutation rate, 70
Germ plasm, constancy of, 94-95
Ground substance of cells, changes in, 62-64
Group participation, 28
Growing Old, 21
Guilt, over aging process, 39

Heredity, and chromosome changes, 91-94
Hierarchic integration concept, 13
Homologous species, 101n
Hormones, 63, 64
Hypoxia, 64-65

I^{131}-albumin, 63
Illness, resilience in recovering from, 40
Immune system, and brain size, 106
Immunology:
  amyloidosis, 82, 85
  antibody production, 81-82
  autoimmune responses, 82
  blood group agglutinogens, 82
  changes in status, 81-83
  gamma globulin, 81
  implications of immunogenetic hypothesis, 84-86
  iso-agglutinin titers, 82
  mitotic inhibition, 85-86
  parabiosis experiments, 85
  X-irradiation, 84, 85
Individual development, pathways of, 13-17

Instrumentality, dominance of, 24
Integration, as adaptive technique, 42
Intellectual functions, and chromosome changes, 92-93
Interiority, focus upon, 26-30
International Gerontological Research Seminar, Sweden, 19n, 21
Interpersonal theory, 37-43
Involvement, as developmental event, 30-32
Iso-agglutinin titers, 82

Kansas City Studies, 21, 24, 25, 28, 32
Kastenbaum, R., 10
Kibbutzim, reading in, 27
King Lear, 107
Klinefelter's syndrome, 95, 96
Knowledge-to-chaos ratio, in learning theory, 45-47

Lansing, A. I., 59
Learning theory:
  Dow theory, 52
  functionalistic ethic, 47-49
  import of gerontology for, 51-53
  knowledge-to-chaos ratio, 45-47
  "meaningful" questions, 46
  nature of learning process, 47-49
  organismic theories, 51-52
  sampling theory, 48-49
Leisure, 19
Life span:
  and brain structure, 99-109
  cell theory, 57-66
  chromosome changes, 91-93
Life span-brain weight equation, 102
"Life-Gestalt," 10, 11
Lifestyle, differentiations in, 22
Lipofuscin, 59, 60
Liver cells, 59, 71-75
Longevity, as organized behavior, 99-

109 *see also* Life span
Loss, reaction to, 40
Lymphocytes, 87-88
Lysosomes, 60
Lytic enzymes, 60

McFarland, 50
Mature individual, image of, 14
Mechanisms of aging, 94-97
Membrane systems, in cells, 61-62
Metals, accumulation in cells, 59
Microenvironmental changes, in cells, 62-64
Microgenesis, 6
Mitochondria, 61
Mitotic errors, 87-91
Mitotic inhibition, 85-86
Molecular biology, postulates of, 99-100
Molecules, and cellular events, 58-62
Mongolism, 78, 94-95
Multiple personality, as solution, 16
Mutation theory *see* Somatic mutation theory

Natural selection, and aging, 95
"Negativism," classic periods of, 31-32
Nervous system *see also* Brain
  chromosomal aberrations, 96
  cognitive role, 106-107
  proteins, 65
  uncertainty factor, 100-103
Neurons, and brain weight, 105-106
Neutron irradiation, 73, 74

Oocytes, 78-79
Organismic theories, 51-52
Organizational participation, 28
Orthogenetic principle, 13

Parabiosis experiments, 85

Paranoid retreat, as adaptive technique, 42
Perivascular glia, effect of age on, 65
Perspective and engrossment concepts, 7-13
Phenotype, term, 101n
Pleiotropism, 96-97
Polyribosomes, 61
Protein, cross-linkages of, 59
Protein molecules, 61-62
Protein synthesis, 60
Proteins, of nervous system, 65
Psychological sequelae, of aging period, 40

RNA, 61, 75-76, 96
Re-engagement, 24
Regression, concept of, 15
Re-organizers of roles, 24
Retirement, 39-40
Ribosomes, 61
Role theory, 15
Rotifera, accumulation of lipofuscin in, 59

Sampling theory, 48-49
Schizophrenia, 30
Self-esteem:
  and activity, 41
  paranoid retreat, 42
  sense of, 37
  threats to, 38-40, 41
"Selfish" periods, of adolescence, 31-32
Sensory disabilities, 40
Shakespeare, 107
Shock, Dr. Nathan, 58
Skinner, 46
Socio-emotionality, dominance of, 24
Somatic mutation theory:

basic theory, 69-70
cancer, 77
chromosome aberrations, 70-75
chromosome healing, 76-77
DNA, 75-76
experimental approach, 70-75
gamma irradiation, 72, 73, 78
gene mutation rate, 70
liver cells, 71-75
neutron irradiation, 73, 74
oocytes, 78-79
problems of theory, 69-70
RNA, 75-76
spermatogenesis, 79
Species, term, 101n
Species phenotypes, stability of, 101
Spermatogenesis, 79
Stress, resilience in recovering from, 40
Subjective time-fields concept, 6
Suicidal tendencies, 42
Sullivan, Harry Stack, 37
Superego, restraints of, 29

Survival, and intellectual performance, 93-94

Television, 28
Time perspective, 6, 8-10
Tokophrya, 59
Twins, 91-93, 96

Uncertainty factor, in performance of living systems, 100-103

Water, in cells, 63
Webster's New International Dictionary, 19
Wechsler-Bellevue scores, and survival, 93
White blood cells, 87-88
Work tasks, meaning of, 23-24

X-irradiation, 84, 85

Youth, worship of, 39

MIX
Papier aus verantwortungsvollen Quellen
Paper from responsible sources
FSC® C105338

If you have any concerns about our products,
you can contact us on
ProductSafety@springernature.com

In case Publisher is established outside the EU,
the EU authorized representative is:
**Springer Nature Customer Service Center GmbH
Europaplatz 3, 69115 Heidelberg, Germany**

Printed by Libri Plureos GmbH
in Hamburg, Germany